MEP 805B / 815B
Generator Set Repair Parts Manual
TM 9-6115-671-24P

Generator Set
Skid Mounted, Tactical Quiet

edited by
Brian Greul

The MEP series of Military Generators are rugged, durable and incorporate proven diesel engine technology. This book is the generator set repair parts manual and also incorporates general support instructions. It is being republished to assist enthusiasts, restorers, and aftermarket owners who use or wish to use these generators outside of military use.

An 8.5x11 3 hole punched loose leaf copy may be purchased for your 3 ring binder. Email books@ocotillopress.com for current information.

Should you have suggestions or feedback on ways to improve this book please send email to Books@OcotilloPress.com

Edited 2021 Ocotillo Press
ISBN 978-1-954285-22-4

Printed in the United States of America

Ocotillo Press
Houston, TX 77017
Books@OcotilloPress.com

Disclaimer: The user of this book is responsible for following safe and lawful practices at all times. The publisher assumes no responsibility for the use of the content of this book. The publisher has made an effort to ensure that the text is complete and properly typeset, however omissions, errors, and other issues may exist that the publisher is unaware of.

TECHNICAL MANUAL

UNIT, DIRECT SUPPORT AND GENER-
AL SUPPORT MAINTENANCE REPAIR PARTS
AND SPECIAL TOOLS LIST

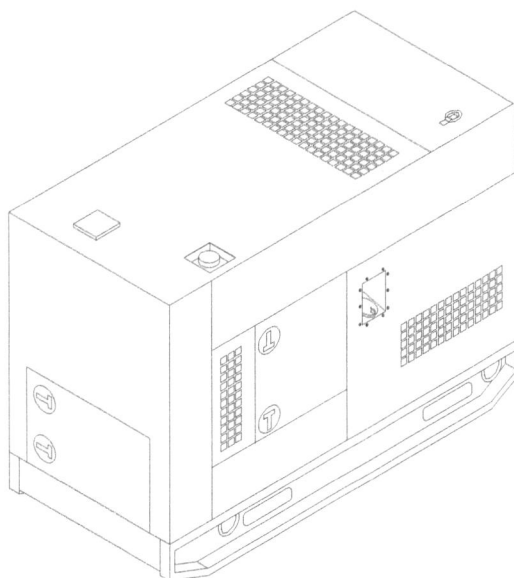

GENERATOR SET,
SKID MOUNTED, TACTICAL QUIET

30 KW, 50/60 AND 400 HZ
MEP-805B (50/60 HZ) (NSN 6115-01-461-9335) (EIC: GGU)
MEP-815B (400 HZ) (NSN 6115-01-462-0290)(EIC: GGV)

Approved for public release; distribution is unlimited.

DEPARTMENTS OF THE ARMY AND THE AIR FORCE
AND HEADQUARTERS, MARINE CORPS
15 AUGUST 2000

LIST OF EFFECTIVE PAGES

NOTE: The portion of the text affected by the changes is indicated by a vertical line in the outer margins of the page. Changes to illustrations are indicated by miniature pointing hands. Changes to wiring diagrams are indicated by shaded areas.

Dates of issue for original and changed pages are:

Original..........0.............15 Aug 2000

Page * Change Page * Change Page * Change No.   No. No. No. No. No.

Page	Change No.	Page	Change No.	Page	Change No.
Title	0	9-1	0	18-1	0
A/(B blank)	0	Fig 10 (3 sheets)	0	Fig 19 (4 sheets)	0
i – xiv	0	10-1 – 10-2	0	19-1	0
Fig. 1 (2 sheets)	0	Fig 11 (3 sheets)	0	Fig 20 (1 sheet)	0
1-1	0	11-1 – 11-2	0	20-1	0
Fig. 2 (2 sheets)	0	Fig 12 (1 sheet)	0	Fig 21 (2 sheets)	0
2-1	0	12-1	0	21-2 – 21-2	0
Fig 3 (1 sheet)	0	Fig 13 (1 sheet)	0	Fig 22 (1 sheet)	0
3-1	0	13-1	0	22-1	0
Fig 4 (10 sheets)	0	Fig 14 (3 sheets)	0	Fig 23 (1 sheet)	0
4-1 – 4-4	0	14-1 – 14-2	0	23-1	0
Fig 5 (11 sheets)	0	Fig 15 (3 sheets)	0	Fig 24 (1 sheet)	0
5-1 – 5-3	0	15-1	0	24-1	0
Fig 6 (3 sheets)	0	Fig 16 (1 sheet)	0	Fig 25 (1 sheet)	0
6-1	0	16-1	0	25-1	0
Fig 7 (3 sheets)	0	Fig 17 (1 sheet)	0	99-1	0
7-1 – 7-2	0	17-1	0	I-1 – I-23	0
Fig 8 (1 sheet)	0	Fig 18 (1 sheet)	0	I-24 blank	0
8-1	0				
Fig 9 (1 sheet)	0				

* Zero in this column indicates an original page.

Technical Manual　　　　　　　　DEPARTMENTS OF THE ARMY No. 9-6115-671-24
　　　　AND THE AIR FORCE, AND

Technical Order　　　　　　　　　HEADQUARTERS, MARINE CORPS No.
35C-3-446-34　　　　　　　　　WASHINGTON D.C., 15 AUGUST 2000 Technical Manual
No. 09249A/09246A-24

UNIT, DIRECT SUPPORT, AND GENERAL SUPPORT MAINTENANCE REPAIR PARTS AND SPECIAL TOOLS LIST

GENERATOR SET, SKID MOUNTED, TACTICAL QUIET 30 KW, 50/60 HZ AND 400 HZ

MEP-805B (50/60 HZ) (NSN 6115-01-461-9335) (EIC: GGU)
MEP-815B (400 HZ) (NSN 6115-01-462-0290) (EIC: GGV)

REPORTING ERRORS AND RECOMMENDING IMPROVEMENTS

You can help improve this manual. If you find any mistakes or if you know of a way to improve the   procedures, please let us know. Mail your letter, DA Form 2028 (Recommended Changes to Publications   and Blank Forms), or DA 2028-2 located in the back of this manual, directly to: Commander, US Army Communications-Electronics Command and Fort Monmouth, ATTN: AMSEL-LC-LEO-D- CS- CFO, Fort Monmouth, New Jersey 07703-5006. The fax number is 732-532-1413, DSN 992-1413. You may also e-mail your recommendations to AMSEL-LC-LEO-PUBS-CHG@mail1. monmouth.army.mil

For Air Force, submit AFTO Form 22 (Technical Order System Publication Improvement Report   and Reply) in accordance with paragraph 6-5, Section VI, TO 00-5-1. Forward direct to prime   ALC/MST.

Marine Corps units, submit NAVMC 10772 (Recommended Changes to Technical Publications) to: Commanding General, Marine Corps Logistics Base (Code 850), Albany, Georgia 31704-5000.

In any case, we will send you a reply.

TABLE OF CONTENTS

TABLE OF CONTENTS - continued

TABLE OF CONTENTS - continued

TABLE OF CONTENTS - continued

TABLE OF CONTENTS - continued

TABLE OF CONTENTS - continued

TABLE OF CONTENTS - continued

TABLE OF CONTENTS - continued

Group / Figure	Title	Page

UNIT, DIRECT SUPPORT, AND
GENERAL SUPPORT MAINTENANCE
REPAIR PARTS AND SPECIAL TOOLS LIST

SECTION 1. INTRODUCTION

1. SCOPE.

This RPSTL lists and authorizes spares and repair parts; special tools; special test, measurement, and diagnostic equipment (TMDE); and other special support equipment required for performance of unit, direct support, and general support maintenance of the generator set. It authorizes the requisitioning, issue, and disposition of spares, repair parts, and special tools as indicated by the source, maintenance and recoverability (SMR) codes.

2. GENERAL.

In addition to Section I, Introduction, this Repair Parts and Special Tools List is divided into the following sections:

    a. <u>Section II. Repair Parts List. A list of spares and repair parts authorized by this RPSTL for use in the</u> performance of maintenance. The list also includes parts which must be removed for replacement of the authorized parts. Parts lists are composed of functional groups in ascending figure and item number sequence. Bulk materials are listed in item name sequence. Repair parts kits are listed separately in their own functional group within Section II. Repair parts for repairable special tools are also listed in this section. Items listed are shown on the associated illustration(s)/figure(s).

    b. <u>Section III. Special Tools List. A list of special tools, special TMDE, and other special support equipment</u> authorized by this RPSTL (as indicated by Basis of Issue (BOI) information in DESCRIPTION AND USABLE ON CODE column) for the performance of maintenance.

    c. <u>Section IV. National Stock Number and Part Number Index. A list, in National Item Identification Number</u> (NIIN) sequence, of all national stock numbered items appearing in the listing, followed by a list in alphanumeric sequence of all part numbers appearing in the listings. National stock numbers and part numbers are cross-referenced to each illustration figure and item number appearance.

3. EXPLANATION OF COLUMNS (SECTIONS II AND III).

    a. <u>ITEM NO, (Column - (1)) . Indicates the number used to identify items called out in the illustration.</u>

b. SMR CODE (Column (2)). The Source, Maintenance, and Recoverability (SMR) code is a 5-position code containing supply/requisitioning information, maintenance category authorization criteria, and disposition instruction, as shown in the following breakout:

Source	Maintenance	Recoverability Code

Source Code   Maintenance   Recoverability Code

Code   Code ☐ ☐_ ☐_ _____   1st two   _____   _____   XX

positions   XX   X ☐   ☐ ☐_

How you get an item   3rd position   4th position   Who determined ☐ ☐   disposition action   tion   Who can install,   Who can do   on an unserviceable   replace or use   complete repair* item.   the item.   on the item.   on the item.

* Complete Repair: Maintenance capacity, capability, and authority to perform all corrective maintenance tasks of the "Repair" function in a use/user environment in order to restore serviceability to a failed item.

(1) Source Code. The source code tells you how to get an item needed for maintenance, repair, or overhaul of an end item/equipment. Explanations of source codes follows:

Code	Explanation

PA
PB   Stocked items; use the applicable NSN to request/requisition items with these source codes. PC** They are authorized to the category indicated by the code entered in the 3rd position
PD   of the SMR code.
PE
PF
PG   ** NOTE: Items coded PC are subject to deterioration.

Code	Explanation

KD   Items with these codes are not to be requested/requisitioned individually. They are part of a KF   kit which is authorized to the maintenance category indicated in the 3rd position of the SMR KB   code. The complete kit must be requisitioned and applied.

Code	Explanation

MO - (Made at unit)   Items with these codes are not to be requested/requisitioned individ- MF - (Made at DS or intermediate)   ually. They must be made from bulk material which is identified MH - (Made at GS)   by the part number in the DESCRIPTION AND USABLE ON CODE ML - (Made at Specialized Repair   (UOC) column and listed in the Bulk Material group of the repair   Activity (SRA))   parts list in this RPSTL. If the item is authorized to you by the 3rd MD - (Made at Depot)   3rd position code of the SMR code, but the source code indicates it is

made at a higher category, order the item from the higher category.

Code   Explanation

AO - (Assembled by unit) Items with these codes are not to be requested/requisitioned   AF - (Assembled by DS or   individually. The parts that make up the assembled item must be

     intermediate) requisitioned or fabricated and assembled at the category of AH - (Assembled by GS) maintenance indicated by the source code. If the 3rd position code of AL - (Assembled by SRA) the SMR authorizes you to replace the item, but the source code   AD - (Assembled by Depot) indicates the item is assembled at a higher category, order the item

         from the higher category of maintenance.

Code   Explanation

XA -   Do not requisition an "XA"-coded item. order its next higher assembly. (Also, refer to the
         NOTE below.)
XB - If an "XB" item is not available from salvage, order it using the FSCM and part number given.   XC - Installation drawing, diagram, instruction sheet, field service drawing, that is identified by
         manufacturer's part number.
XD - Item is not stocked. Order an "XD"-coded item through normal supply channels using the
         FSCM and part number given, if no NSN is available.

         NOTE: Cannibalization or controlled exchange, when authorized, may be used as a source of supply f or items with the above source codes, except for those source coded "XA" or those aircraft support items restricted by requirements of AR 700-42.

     (2) Maintenance Code. Maintenance codes tell you the category (s) of maintenance authorized to USE and REPAIR support items. The maintenance codes are entered in the 3rd and 4th positions of the SMR Code as follows:

     (a) The maintenance code entered in the 3rd position tells you the lowest maintenance category authorized to remove, replace, and use an item. The maintenance code entered in the 3rd position will indicate authorization to one of the following categories of maintenance.

     Code   Application/Explanation C   Crew or operator mainte-

nance done within unit maintenance.

     O   Unit category can remove, replace, and use the item.

     F   Direct support or intermediate category can remove, replace, and use the item.

     H   General support catagory can remove, replace, and use the item.

     L   Specialized repair activity can remove, replace, and use the item.

     D   Depot catagory can remove, replace, and use the item.

(b) The maintenance code entered in the 4th position tells whether or not the item is to be repaired and identifies the lowest maintenance category with the capability to do complete repair (i.e., perform all authorized repair functions).

> **NOTE**: Some limited repair may be done on the item at a lower category of maintenance, if authorized by the Maintenance Allocation Chart (MAC) and SMR Codes). This position will contain one of the following maintenance codes.

Maintenance
Code   Application/Explanation

O   Unit is the lowest category that can do complete repair of the item.

F   Direct support or intermediate is the lowest category that can do complete repair of the item.

H   General support is the lowest category that can do complete repair of the item.

L   Specialized repair activity (designate the specialized repair activity) is the lowest category that can do complete repair of the item.

D   Depot is the lowest category that can do complete repair of the item.

Z   Nonrepairable. No repair is authorized.

B   No repair is authorized. (No parts or special tools are authorized for the maintenance of a "B" coded item). However, the item may be reconditionedbyadjusting, lubricating, etc., at the user category.

(3) Recoverability Code. Recoverability codes are assigned to items to indicate the disposition action on unserviceable items. The recoverability code is entered in the 5th position of the SMR Code as follows:

Recoverability
Code   Application/Explanation

Z   Nonrepairable item. When unserviceable, condemn and dispose of the item at the category of maintenance shown in 3rd position of SMR Code.

O   Repairable item. When uneconomically repairable, condemn and dispose of the item at unit category.

F   Repairable item. when uneconomically repairable, condemn and dispose of the item at the direct support or intermediate category.

H   Repairable item. when uneconomically repairable, condemn and dispose of the item at the general support category.

Recoverability
Code   Application/Explanation

D   Repairable item. When beyond lower category repair capability, return to depot.
        Condemnation and disposal of item not authorized below depot category.

L   Repairable item. Condemnation and disposal not authorized below specialized repair
        activity (SRA).

A   Item requires special handling or condemnation procedures because of specific reasons
        (e.g., precious metal content, high dollar value, critical material, or hazardous material).
        Refer to appropriate manuals/directives for specific instructions.

    c. CAGEC (Column (3)). The Commercial and Government Entity Code (CAGEC) is a 5-digit numeric code which is used to identify the manufacturer, distributor, or government agency, etc., that supplies the item.

    d. PART NUMBER (Column (4)). Indicates the primary number used by the manufacturer (individual, company" firm, corporation, or Government activity), which controls the design and characteristics of the item by means of its engineering drawings, specifications standards, and inspection requirements to identify an item or range of items.

NOTE: When you use a NSN to requisition an item, the item you receive may have a different part number from the part ordered.

    e. DESCRIPTION AND USABLE ON CODE (UOC) (Column (5)). This column includes the following information:

    (1) The Federal item name and, when required, a minimum description to identify the item.

    (2) Items that are included in kits and sets are listed below the name of the kit or set.

    (3) Spare/repair parts that make up an assembled item are listed immediately following the assembled item line entry.

    (4) Part numbers for bulk materials are referenced in this column in the line item entry for the item to be manufactured/fabricated.

    (5) When the item is not used with all serial numbers of the same model, the effective serial numbers are shown on the last line(s) of the description (before UOC).

    (6) The UOC, when applicable (see paragraph 5, Special Information).

    (7) In the Special Tools List section, the Basis of Issue (BOI) appears as the last line(s) in the entry for each special tool, special TMDE, and other special support equipment. When density of equipment supported exceeds density spread indicated in the BOI, the total authorization is increased proportionately.

    (8) The statement "END OF FIGURE" appears just below the last item description in Column 5 for a given figure in Section II.

f. OTY (Column (6)). The QTY (quantity per figure column) indicates the quantity of the item used in the breakout shown on the illustration figure, which is prepared for a functional group, subfunctional group, or an assembly. A "V" appearing in this column in lieu of a quantity indicates that the quantity is variable and that the quantity may vary from application to application.

g. EXPLANATION OF COLUMNS (SECTION IV).

a. National Stock Number (NSN) Index.

(1) STOCK NUMBER Column. This column lists the NSN by National Item Identification Number (NIIN) sequence. The NIIN consists of the last nine digits of the NSN.

$$\underline{\text{NSN}}$$
$$(5305 - \underline{01 - 674 - 1467})$$
$$\text{NIIN}$$

When using this column to locate an item, ignore the first 4 digits of the NSN. However, the complete
NSN should be used when ordering items by stock number.

(2) FIG. Column. This column lists the number of the figure where the item is identified/located. The figures are in numerical order in Section II and Section III.

(3) ITEM Column. The item number identifies the item associated with the figure listed in the adjacent FIG. Column. This item is also identified by the NSN listed on the same line.

b. Part Number Index. Part numbers in this index are listed by part number in ascending alphanumeric sequence (i.e., vertical arrangement of letter and number combination which places the first letter or digit of each group in order A through Z, followed by numbers 0 through 9 and each following letter or digit in like order).

(1) CAGEC Column. The Commercial and Government Entity Code (CAGEC) is a 5-digit numeric code used to identify the manufacturer, distributor, or Government agency, etc., that supplies the item.

(2) PART NUMBER Column. Indicates the primary number used by the manufacturer (individual, firm, corporation, or Government activity) , which controls the design and characteristics of the item by means of its engineering drawings, specifications standards, and inspection requirements to identify an item or range of items.

(3) STOCK NUMBER Column. This column lists the NSN for the associated part number and manufacturer identified in the PART NUMBER and FSCM columns to the left.

(4) FIG. Column. This column lists the number of the figure where the item is identified/located in Section II and III.

(5) ITEM Column. The item number is that number assigned to the item as it appears in the figure referenced in the adjacent figure number column.

5. SPECIAL INFORMATION. Use the following subparagraphs as applicable:

6. <u>Usable on Code.</u> The usable on code appears in the lower left corner of the Description column heading. Usable on codes are shown as "UOC: . . ." under the applicable item description/nomenclature. Uncoded items are applicable to all models. Identification of the usable on codes used in the RPSTL are:

<u>Code</u>	Used On
LTX	Generator Set, Skid Mounted, Tactical Quiet, 30 KW, 50/60 Hz
LTY	Generator Set, Skid Mounted, Tactical Quiet, 30 KW, 400 Hz

b. <u>Fabrication Instructions.</u> Bulk materials required to manufacture items are listed in the Bulk Material Functional Group of this RPSTL. Part numbers for bulk materials are also referenced in the description column of the line item entry for the item to be manufactured/fabricated. Detailed fabrication instructions for item source codes to be manufactured or fabricated are found in TM 9-6115-671-14.

c. <u>Assembly Instruction.</u> Detailed assembly instructions f or items source coded to be assembled from component spare/repair parts are found in TM 9-6115-671-14. Items that make up the assembly are listed immediately following the assembly item entry or reference is made to an applicable figure.

d. <u>Kits.</u> Line item entries for repair parts kits appear throughout Section II.

e. <u>Associated Publications.</u> The publication(s) listed below pertain to this generator set and its components:

<u>Publication</u>	<u>Short Title</u>
TM 9-6115-671-14	Maintenance Manual, Generator Set, Skid Mounted, Tactical Quiet, 30 KW, 50/60 Hz and 400 Hz

f. <u>National Stock Numbers.</u> National stock numbers (NSN's) that are missing from "P" source coded items have been applied for and will be added to this TM by future change/revision when they are entered in the Army Master Data File (AMDF). Until the NSN's are established and published, submit exception requisitions to: Commander, US Army Communications-Electronics Command and Fort Monmouth, ATTN: AMSEL-LC-MM, Fort Monmouth, NJ 07703-5007 for the part required to support your equipment.

6. HOW TO LOCATE REPAIR PARTS.

a. <u>When National Stock Number or Part Number is Not Known.</u>

(1) <u>First.</u> Using the Table of Contents, determine the assembly group or subassembly group to which the item belongs. This is necessary since figures are prepared for assembly groups and subassembly groups, and listings are divided into the same groups.

(2) <u>Second.</u> Find the figure covering the assembly group or subassembly group to which the item belongs.

(3) <u>Third.</u> Identify the item on the figure and note the item number.

(4) <u>Fourth</u>. Refer to the Repair Parts List for the figure to find the part number for the item number noted on the figure.

(5) <u>Fifth</u>. Refer to the Part Number Index to find the NSN, if assigned.

b. <u>When National Stock Number or Part Number is Known.</u>

(1) <u>First</u>. Using the Index of National Stock Numbers and Part Numbers, find the pertinent National Stock Number or Part Number. The NSN index is in National Item Identification Number (NIIN) sequence (see 4. a (1)) . The part numbers in the Part Number index are listed in ascending alphanumeric sequence (see 4.b). Both indexes cross-reference you to the illustration figure and item number of the item you are looking for.

(2) <u>Second</u>. After finding the figure and item number, verify that the item is the one you're looking for, then locate the item number in the repair parts list for the figure.

7. ABBREVIATIONS.

<u>Abbreviations</u> <u>Explanation</u> BOI   Basis of Issue

DS   Direct Support

GS   General Support

MAC   Maintenance Allocation Chart

NIIN   National Item Identification Number
(consists of the last 9-digits of the NSN)

NSN   National Stock Number

RPSTL   Repair Parts and Special Tools List

SMR   Source, Maintenance and Recoverability Codes SRA   Special Repair Activity

TMDE   Test, Measurement and Diagnostic Equipment UOC   Usable on Code

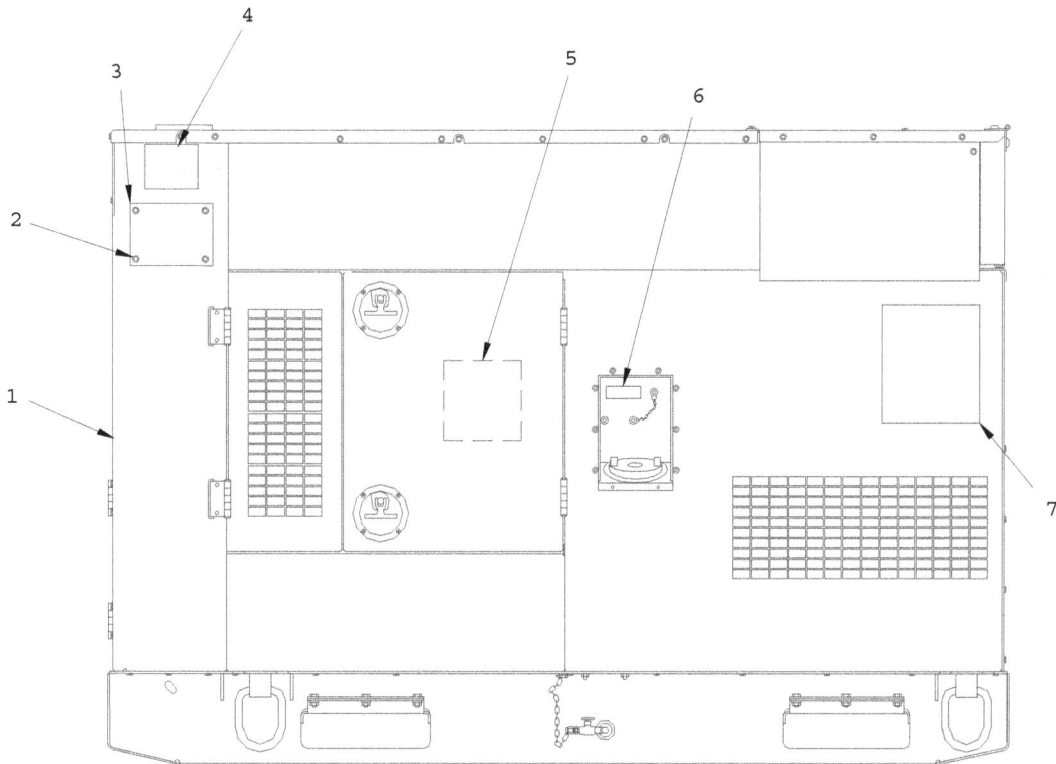

LEFT SIDE

REAR

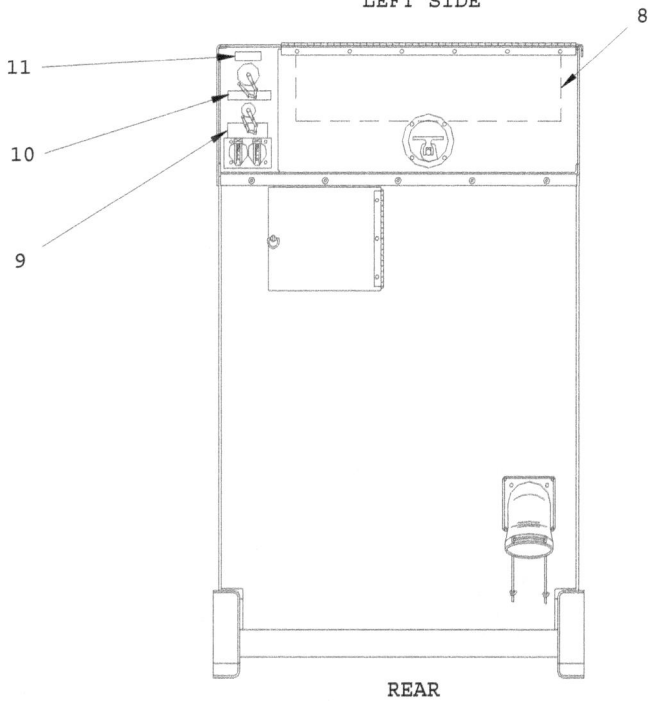

Figure 1. Generator Set, 30 KW (Sheet 1 of 2)

RIGHT SIDE

FRO NT

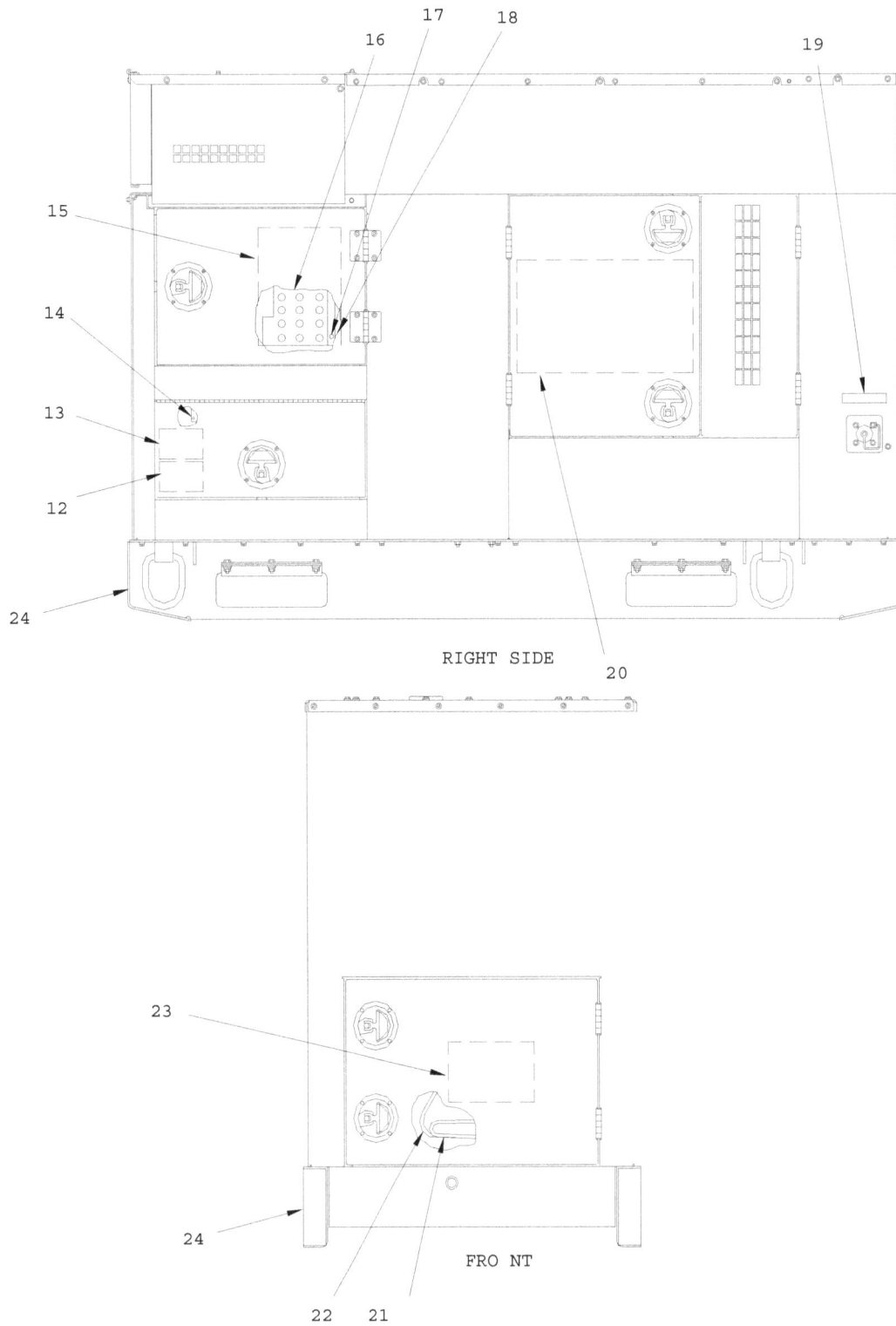

Figure 1. Generator Set, 30 KW (Sheet 2 of 2)

(1)	(2)		(3)	(4) SMR CODE	(5)	(6)
ITEM NO	ARMY	AIR FORCE USMC	CAGEC	PART NUMBER	DESCRIPTION AND USABLE ON CODE (UOC)	QTY

GROUP 00 GENERATOR SET
FIGURE 1 GENERATOR SET

ITEM NO	ARMY	AIR FORCE	USMC	CAGEC	PART NUMBER	DESCRIPTION AND USABLE ON CODE (UOC)	QTY
1	XCOFF	PA000	PBOFF	30554	96-23500	GENERATOR SET ASSEMBLY,50/60 HZ.......1 UOC: LTY	
1	XCOFF	PA000	PBOFF	30554	96-23501	GENERATOR SET ASSEMBLY,400 HZ.........1	
UOC: LTX							
2	XBOZZ	XB	XBOZZ	81349	M24243/6-A402H	RIVET,BLIND.........................70	
3	XBOZZ	XB	XBOZZ	30554	96-23571-01	PLATE,IDENTIFICATION,SET RATING.......1 UOC: LTY	
3	XBOZZ	XB	XBOZZ	30554	96-23571-02	PLATE,IDENTIFICATION,SET RATING.......1	
UOC: LTX							
4	XBOZZ	XB	XBOZZ	30554	96-23506-01	PLATE,IDENTIFICATIO...................1 UOC: LTY	
4	XBOZZ	XB	XBOZZ	30554	96-23506-02	PLATE,IDENTIFICATIO...................1 UOC: LTX	
5	XBOZZ	XB	XBOZZ	30554	88-21634	PLATE,FUEL SYSTEM D...................1	
6	XBOZZ	XB	XBOZZ	30554	88-20102	PLATE,IDENTIFICATIO...................1	
7	XBOZZ	XB	XBOZZ	30554	96-23510	PLATE,INSTRUCTION,LIFTING.............1	
8	XBOZZ	XB	XBOZZ	30554	96-23512	PLATE,INSTRUCTION,OPERATION...........1 UOC: LTY	
8	XBOZZ	XB	XBOZZ	30554	96-23513	PLATE,INSTRUCTION,OPERATION...........1 UOC: LTX	
9	XBOZZ	XB	XBOZZ	30554	88-20073	PLATE,CONVENIENCE RECEPTACLE..........1 UOC: LTY	
9	XBOZZ	XB	XBOZZ	30554	88-22737	PLATE,CONVENIENCE RECEPTACLE..........1 UOC: LTX	
10	XBOZZ	XB	XBOZZ	30554	88-20074	PLATE,PARALLELING RECEPTACLE..........1	
11	XBOZZ	XB	XBOZZ	30554	96-23604	PLATE,IDENTIFICATIO,COMMUNICATION.....1	
12	XBOZZ	XB	XBOZZ	30554	88-20110	PLATE,CAUTION,VOLTAGE CONNECTION......1	
13	XBOZZ	XB	XBOZZ	30554	88-20126	PLATE,GROUNDING STU...................1	
14	XBOZZ	XB	XBOZZ	30554	88-22738	PLATE,SWITCH BOX RECEPTACLE...........1	
15	XBOZZ	XB	XBOZZ	30554	96-23596	PLATE,IDENTIFICATIO,SCHEMATIC.........1	
16	PAOZZ	PA000	PAOZZ	30554	88-22509	COVER,DISTRIBUTION CONNECTION.........1	
17	PAOZZ	PA000	PAOZZ	96906	MS27183-10	WASHER,FLAT...........................4	
18	PAOZZ	PA000	PAOZZ	30554	88-20568-1	NUT,PLAIN,CASTELLAT...................4	
19	XBOZZ	XB	XBOZZ	30554	88-20075	PLATE,SLAVE RECEPTACLE................1	
20	XBOZZ	XB	XBOZZ	30554	96-23509	PLATE,IDENTIFICATIO,WIRING............1	
21	PAOZZ	PA000	PAOZZ	30554	88-22209	CABLE ASSEMBLY,SPE....................1	
22	PAOZZ	PA000	PAOZZ	01276	FA1493FFF3000	HOSE ASSEMBLY,NONME...................1	
23	XBOZZ	XB	XBOZZ	30554	88-21603	PLATE,INSTRUCTION,BATTERY.............1	
24	XBFFF	XB	XBFFF	30554	88-22008	SKID BASE ASSEMBLY....................1 (SEE FIGURE 25 FOR PARTS BREAKDOWN)	

END OF FIGURE

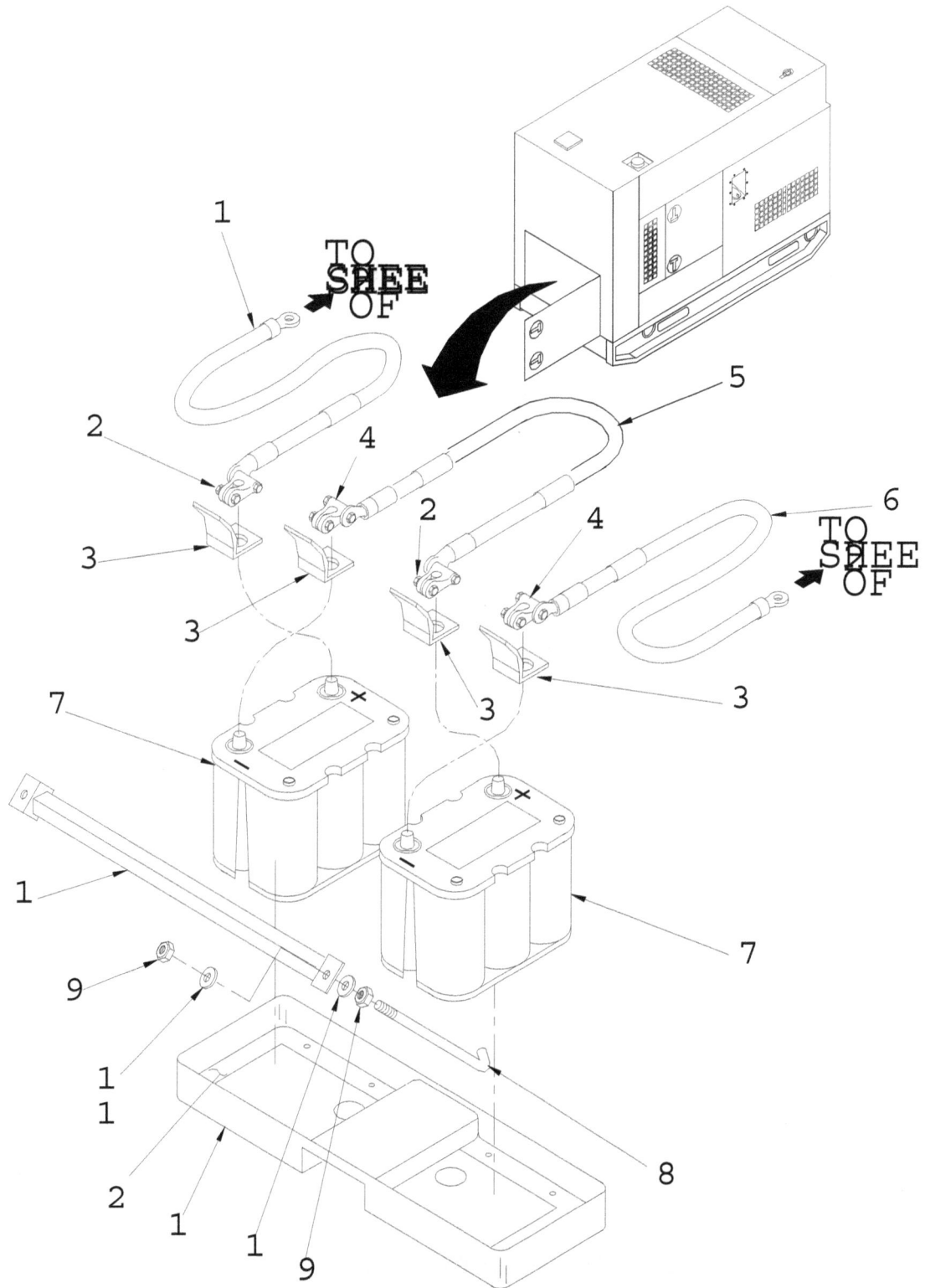

Figure 2. DC Electrical System (Sheet 1 of 2)

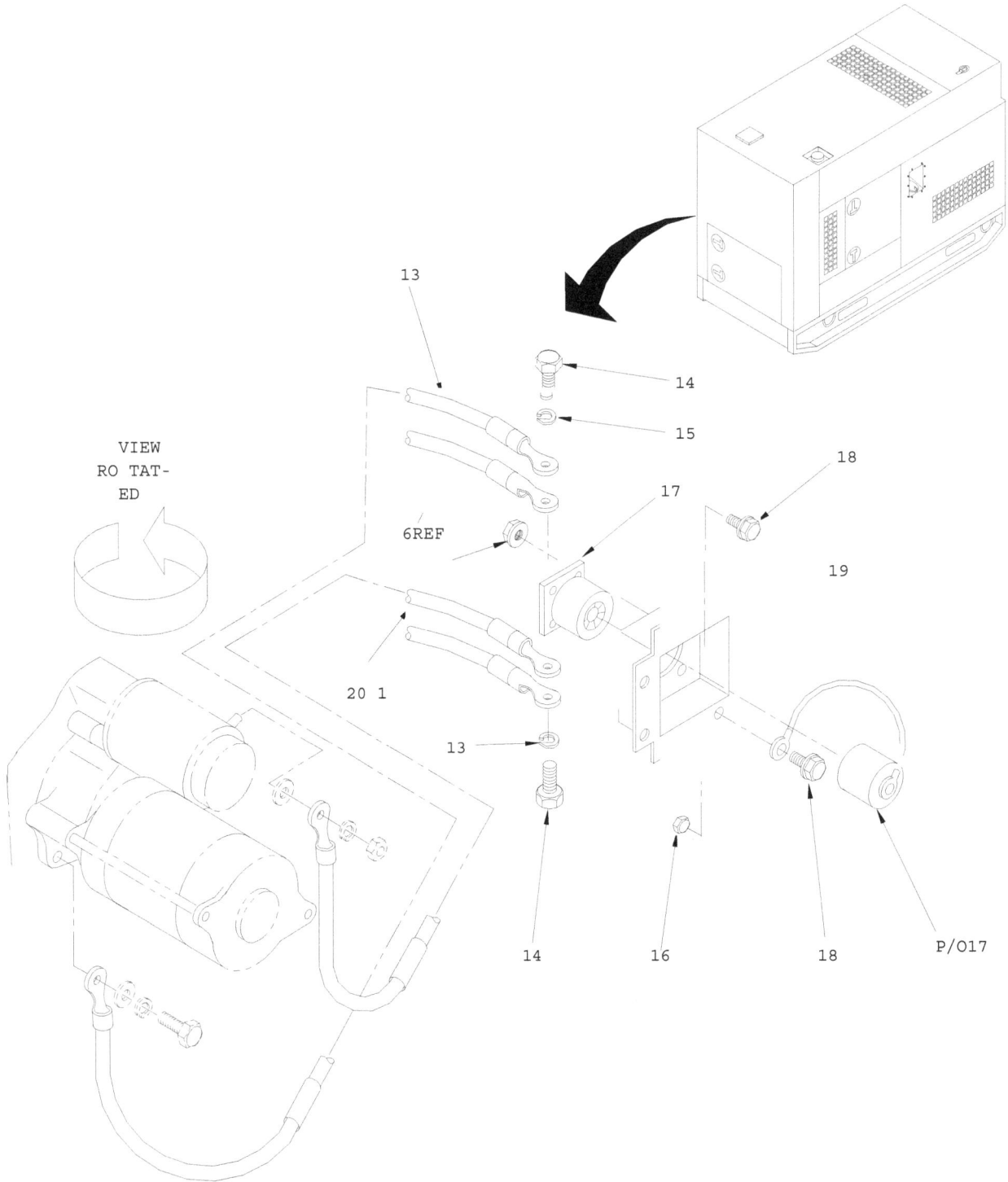

13

14

15

18

17

19

6REF

VIEW
RO TAT-
ED

20 1

13

14

16

18

P/O17

Figure 2. DC Electrical System (Sheet 2 of 2)

(1)		(2)		(3)	(4) SMR CODE	(5)	(6)
ITEM NO	ARMY	AIR FORCE	USMC	CAGEC	PART NUMBER	DESCRIPTION AND USABLE ON CODE (UOC)	QTY

GROUP 01 DC ELECTRICAL SYSTEM
FIGURE 2 DC ELECTRICAL SYSTEM

1	A0000	M000	M0000	30554	96-23563	CABLE ASSEMBLY,BATT....................1
						(SEE FIGURE 3 FOR PARTS BREAKDOWN)
2	PA0ZZ	PA000	PA0ZZ	4U407	A52425-1	TERMINAL,LUG.........................2
3	PA0ZZ	PA000	PA0ZZ	30554	88-20188	CAP,ELECTRICAL......................4
4	PA0ZZ	PA000	PA0ZZ	4U407	A52425-2	TERMINAL,LUG.........................2
5	A0000	M000	M0000	30554	96-23561	CABLE ASSEMBLY,BATT....................1
						(SEE FIGURE 3 FOR PARTS BREAKDOWN)
6	A0000	M000	M0000	30554	96-23562	CABLE ASSEMBLY,BATT....................1
						(SEE FIGURE 3 FOR PARTS BREAKDOWN)
7	PA0ZZ	PA000	PA0ZZ	0UJ55	800S	BATTERY,STORAGE......................2
8	XB0ZZ	XB	XB0ZZ	30554	88-21685	BOLT,HOOK...........................2
9	PA0ZZ	PA000	PA0ZZ	30554	88-20568-2	NUT,SELF-LOCKING....................4
10	XB0ZZ	XB	XB0ZZ	30554	88-21683	TRAY,BATTERY.........................1
11	PAFZZ	PA000	PAFZZ	96906	MS21306-1G	WASHER,FLAT...........................4
12	XB0ZZ	XB	XB0ZZ	30554	88-21684	HOLDDOWN,BATTERY......................1
13	A0000	M000	M0000	30554	96-23564	CABLE ASSEMBLY,BATT....................1
						(SEE FIGURE 3 FOR PARTS BREAKDOWN)
14	PA0ZZ	PA000	PA0ZZ	80204	B1821BH038C063N	SCREW,CAP,HEXAGON H..................2
15	PA0ZZ	PA000	PA0ZZ	97403	13230E6744-65	WASHER,LOCK..........................4
16	PA0ZZ	PA000	PA0ZZ	19617	HW310	NUT,PLAIN,ASSEMBLED..................12
17	PA0ZZ	PA000	PA0ZZ	19207	11674728	CONNECTOR,RECEPTACA..................1
18	PA0ZZ	PA000	PA0ZZ	19207	12325869	BOLT,MACHINE........................12
19	XB0ZZ	XB	XB0ZZ	30554	88-21952	BOX,SLAVE RECEPTACL..................1
20	A0000	M000	M0000	30554	96-23566	CABLE ASSEMBLY,BATT....................1
						(SEE FIGURE 3 FOR PARTS BREAKDOWN)
21	XB0ZZ	XB	XB0ZZ	30554	96-23654	ADAPTER,BATTERY TRA..................2

END OF FIGURE

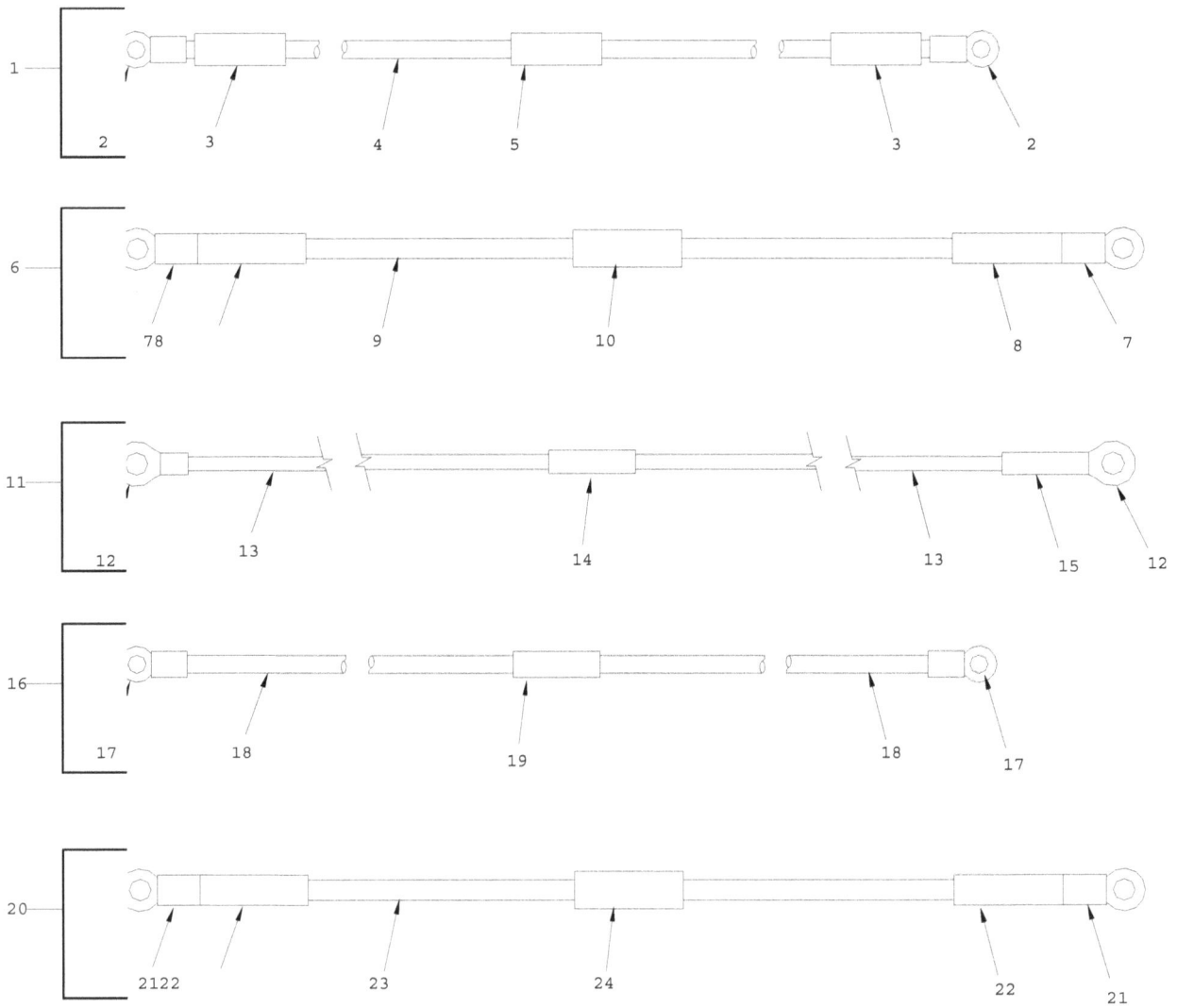

Figure 3. Battery and Slave Receptacle Cables

(1) ITEM NO	(2) ARMY	AIR FORCE	USMC	(3) CAGEC	(4) SMR CODE PART NUMBER	(5) DESCRIPTION AND USABLE ON CODE (UOC)	(6) QTY
						GROUP 0102 BATTERY AND SLAVE RECEPTACLE CABLES	
						FIGURE 3 BATTERY AND SLAVE RECEPTACLE CABLES	
1	A0000	M000	M0000	30554	96-23563	CABLE ASSEMBLY,BATT....................1	
2	PA0ZZ	PA000	PA0ZZ	98410	H-780-38	TERMINAL,LUG............................2	
3	PA0ZZ	PA000	PA0ZZ	28105	ST-301-1RED	INSULATION,SLEEVING....................V	
4	M00ZZ	M000	M0000	30554	96-23637	CABLE,ELECTRICAL.......................V	
						MAKE FROM 1/0SGT (5G996), 15.0 IN	
5	PA0ZZ	PA000	PA0ZZ	28105	ST-301-1WHITE	INSULATION,SLEEVING....................V	
6	A0000	M000	M0000	30554	96-23566	CABLE ASSEMBLY,BATT....................1	
7	PA0ZZ	PA000	PA0ZZ	98410	H-780-38	TERMINAL,LUG............................2	
8	PA0ZZ	PA000	PA0ZZ	28105	ST-301-1RED	INSULATION,SLEEVING....................V	
9	M00ZZ	M000	M0000	30554	96-23637	CABLE,ELECTRICAL.......................V	
						MAKE FROM 1/0SGT (5G996), 48.0 IN	
10	PA0ZZ	PA000	PA0ZZ	28105	ST-301-1WHITE	INSULATION,SLEEVING....................V	
11	A0000	M000	M0000	30554	96-23562	CABLE ASSEMBLY,BATT....................1	
12	PA0ZZ	PA000	PA0ZZ	98410	H-780-38	TERMINAL,LUG............................2	
13	M00ZZ	M000	M0000	30554	96-23637	CABLE,ELECTRICAL.......................1	
						MAKE FROM 1/0SGT (5G996), 47.0 IN	
14	PA0ZZ	PA000	PA0ZZ	28105	ST-301-1WHITE	INSULATION,SLEEVING....................V	
15	PA0ZZ	PA000	PA0ZZ	28105	ST-301-1RED	INSULATION,SLEEVING....................V	
16	A0000	M000	M0000	30554	96-23564	CABLE ASSEMBLY,BATT....................1	
17	PA0ZZ	PA000	PA0ZZ	98410	H-780-38	TERMINAL,LUG............................2	
18	M00ZZ	M000	M0000	30554	96-23637	CABLE,ELECTRICAL.......................V	
						MAKE FROM 1/0SGT (5G996), 50.0 IN	
19	PA0ZZ	PA000	PA0ZZ	28105	ST-301-1WHITE	INSULATION,SLEEVING....................V	
20	A0000	M000	M0000	30554	96-23561	CABLE ASSEMBLY,BATT....................1	
21	PA0ZZ	PA000	PA0ZZ	98410	H-780-38	TERMINAL,LUG............................2	
22	M00ZZ	M000	M0000	30554	96-23637	CABLE,ELECTRICAL.......................V	
						MAKE FROM 1/0SGT (5G996), 14.0 IN	
23	PA0ZZ	PA000	PA0ZZ	28105	ST-301-1RED	INSULATION,SLEEVING....................V	
24	PA0ZZ	PA000	PA0ZZ	28105	ST-301-1WHITE	INSULATION,SLEEVING....................V	

END OF FIGURE

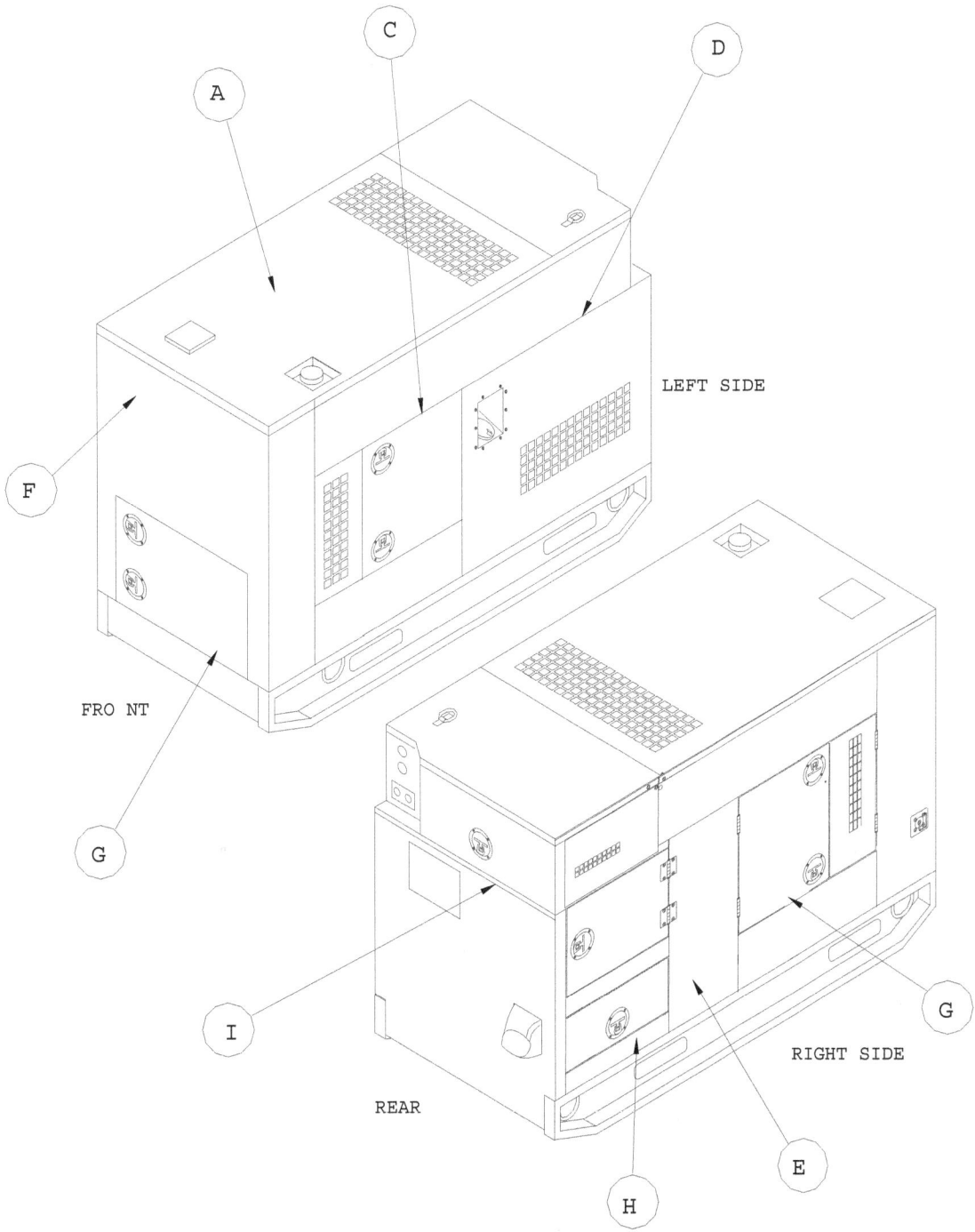

Figure 4. Housing (Sheet 1 of 10)

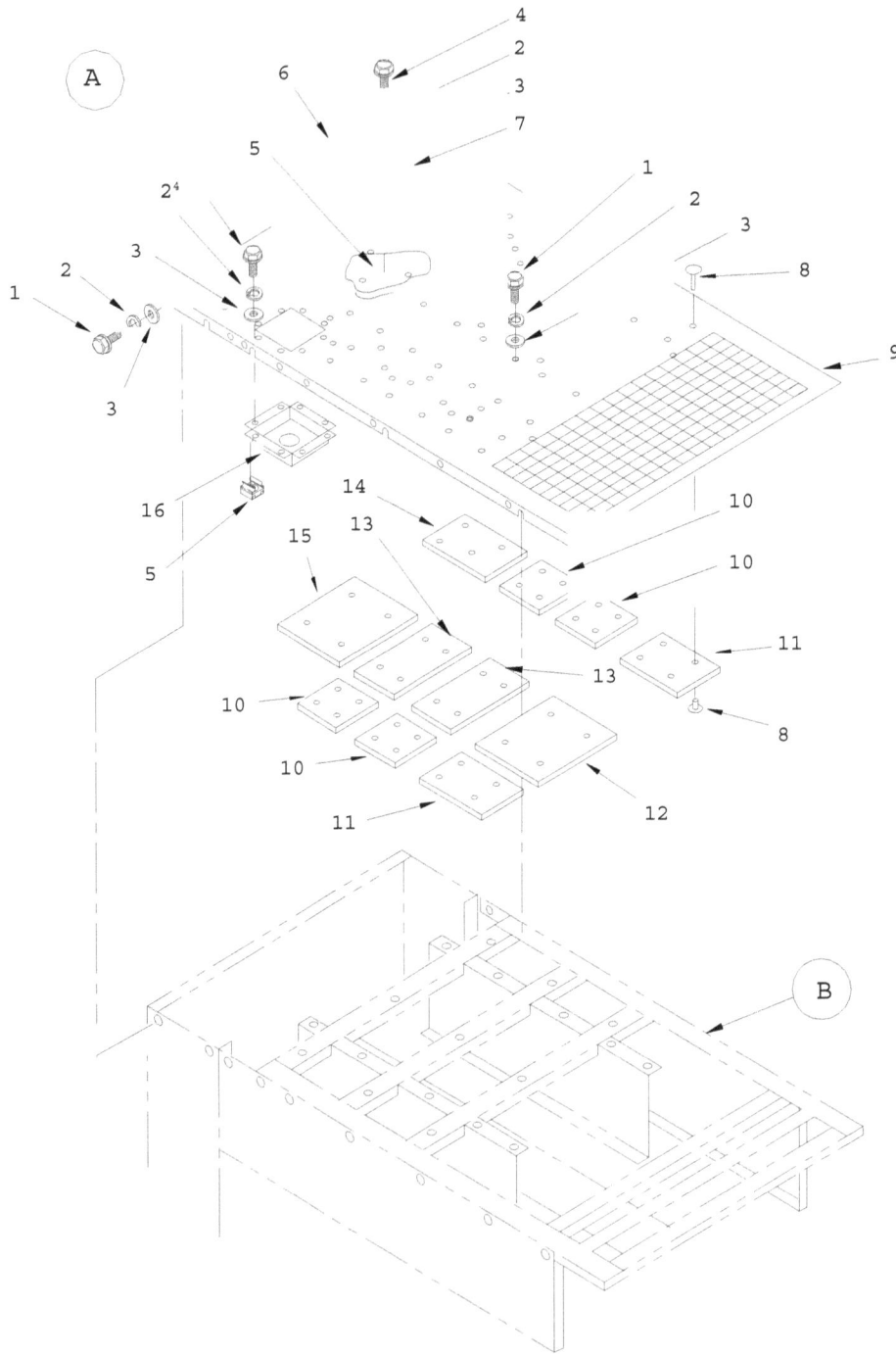

Figure 4. Housing (Sheet 2 of 10)

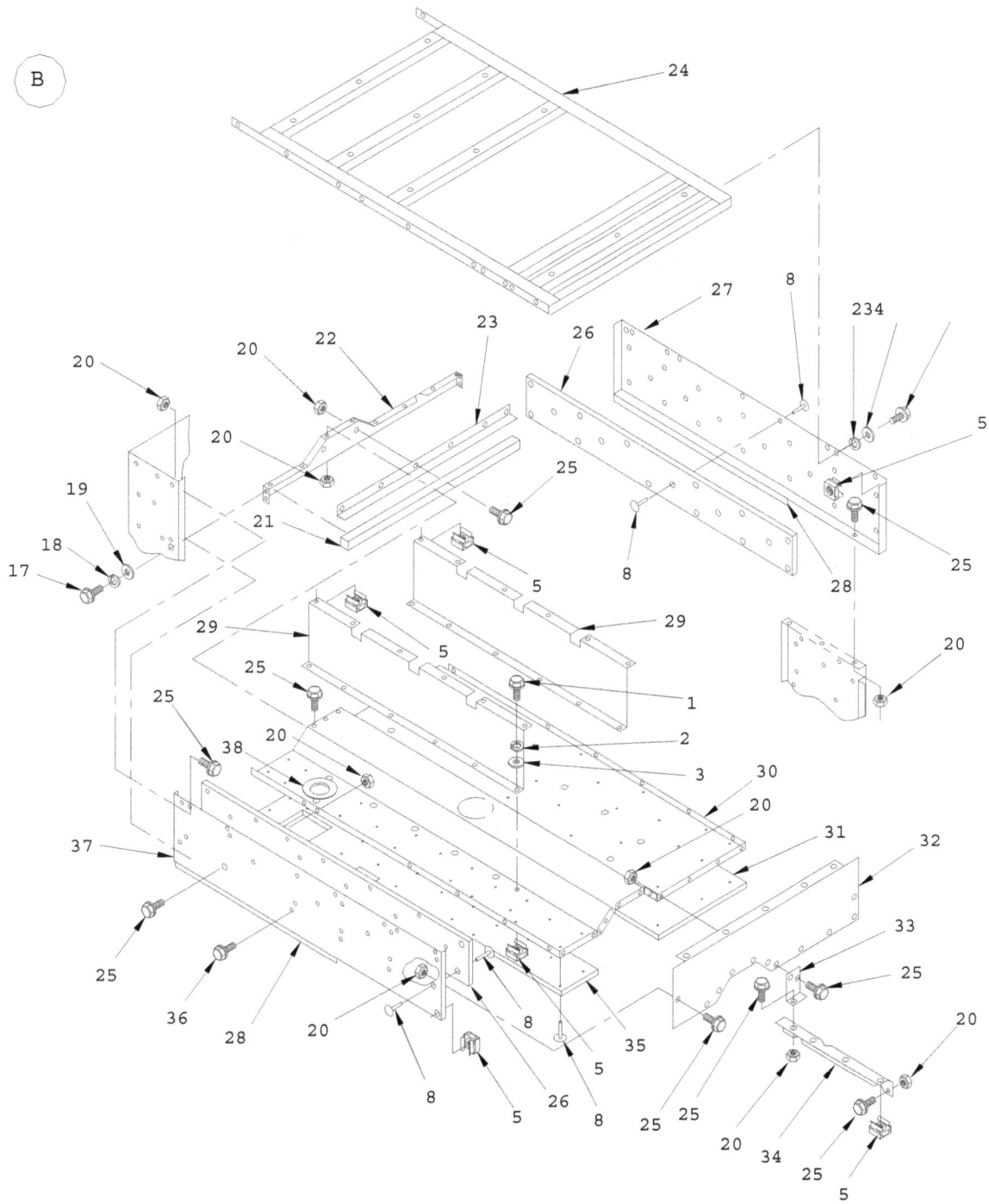

Figure 4. Housing (Sheet 3 of 10)

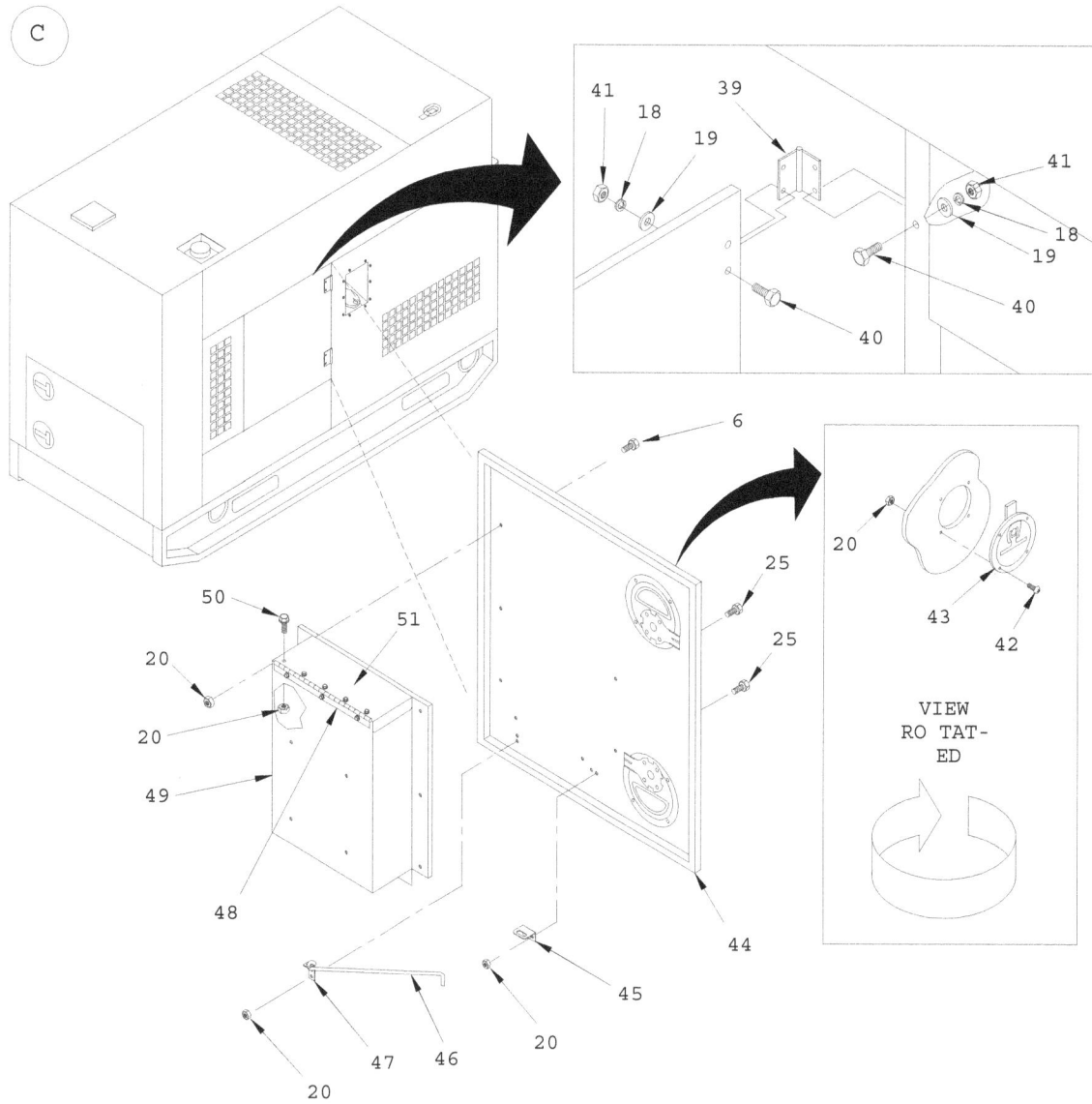

Figure 4. Housing (Sheet 4 of 10)

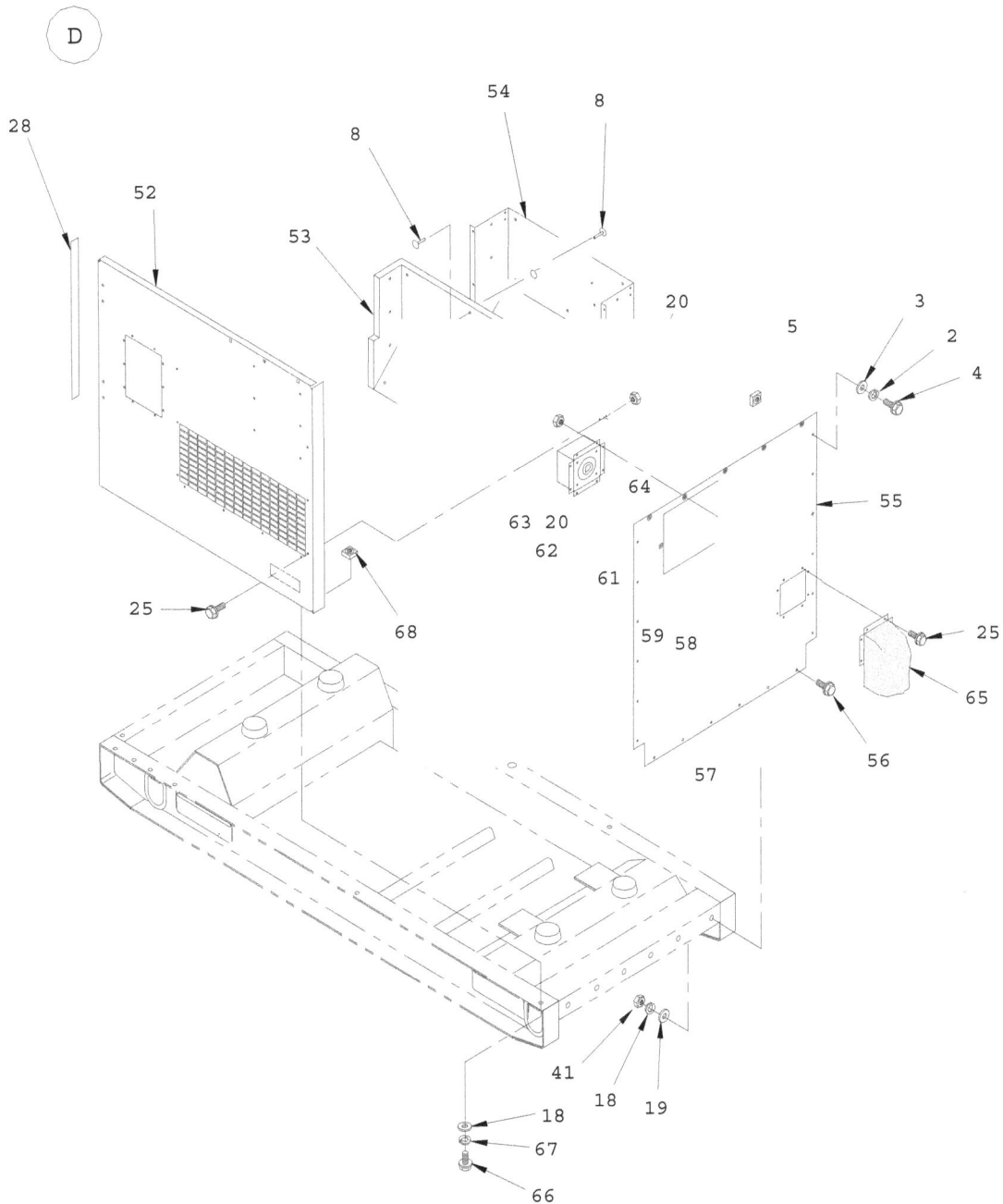

Figure 4. Housing (Sheet 5 of 10)

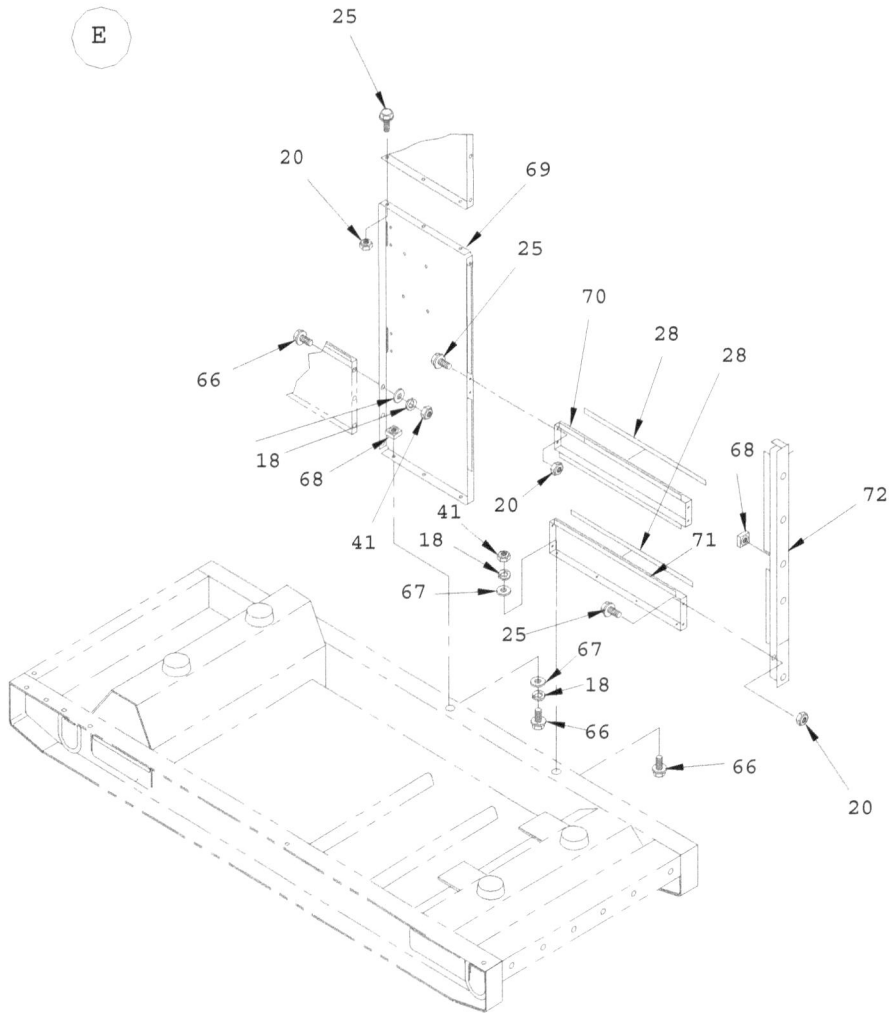

Figure 4. Housing (Sheet 6 of 10)

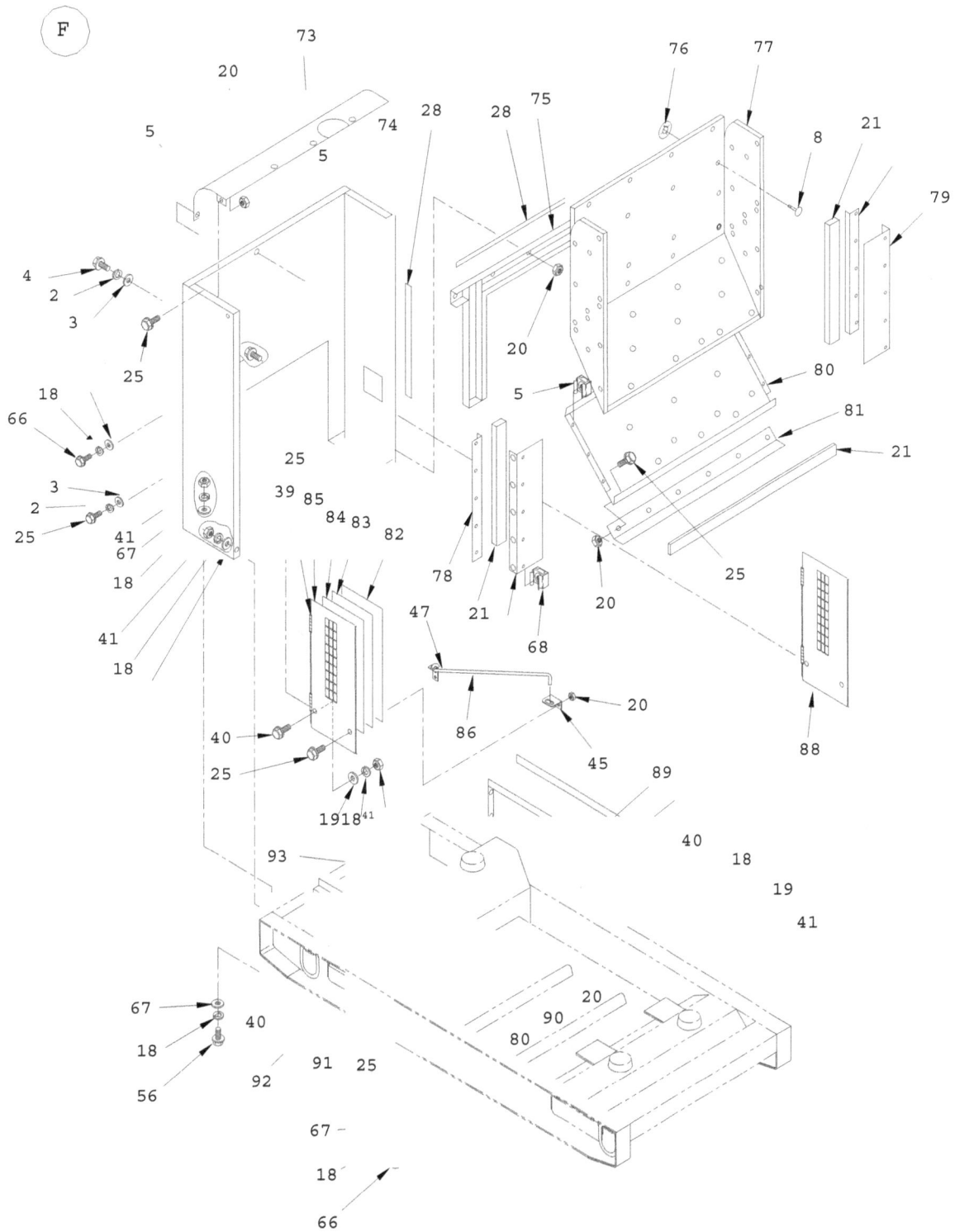

Figure 4. Housing (Sheet 7 of 10)

G

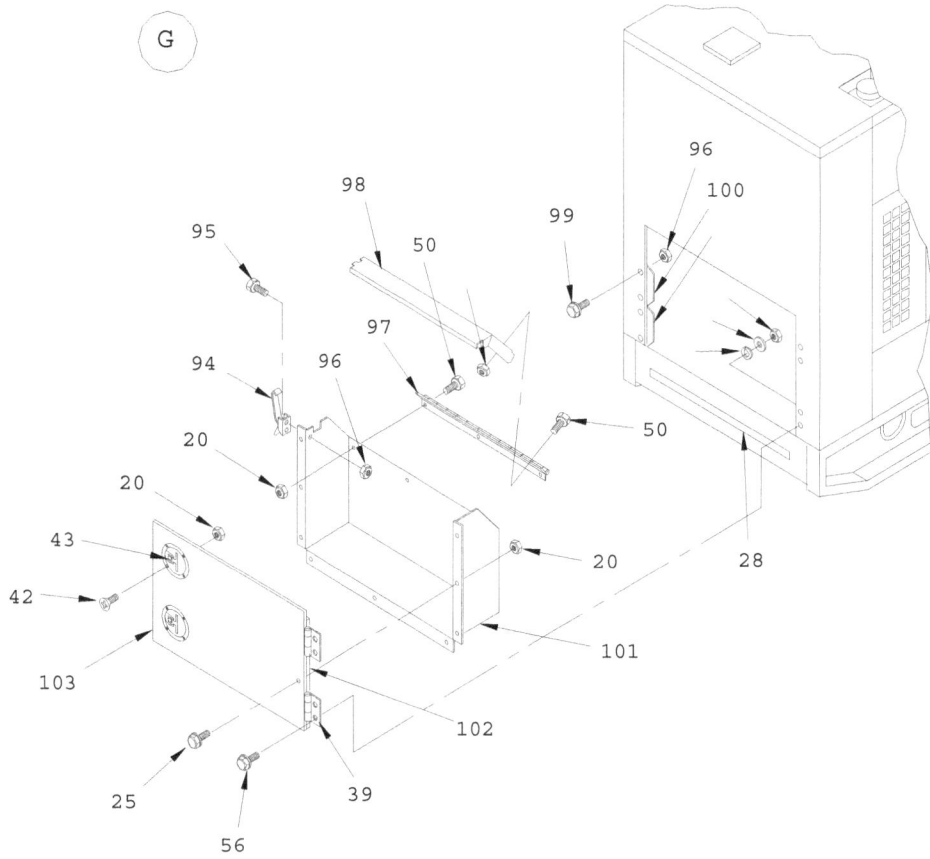

Figure 4. Housing (Sheet 8 of 10)

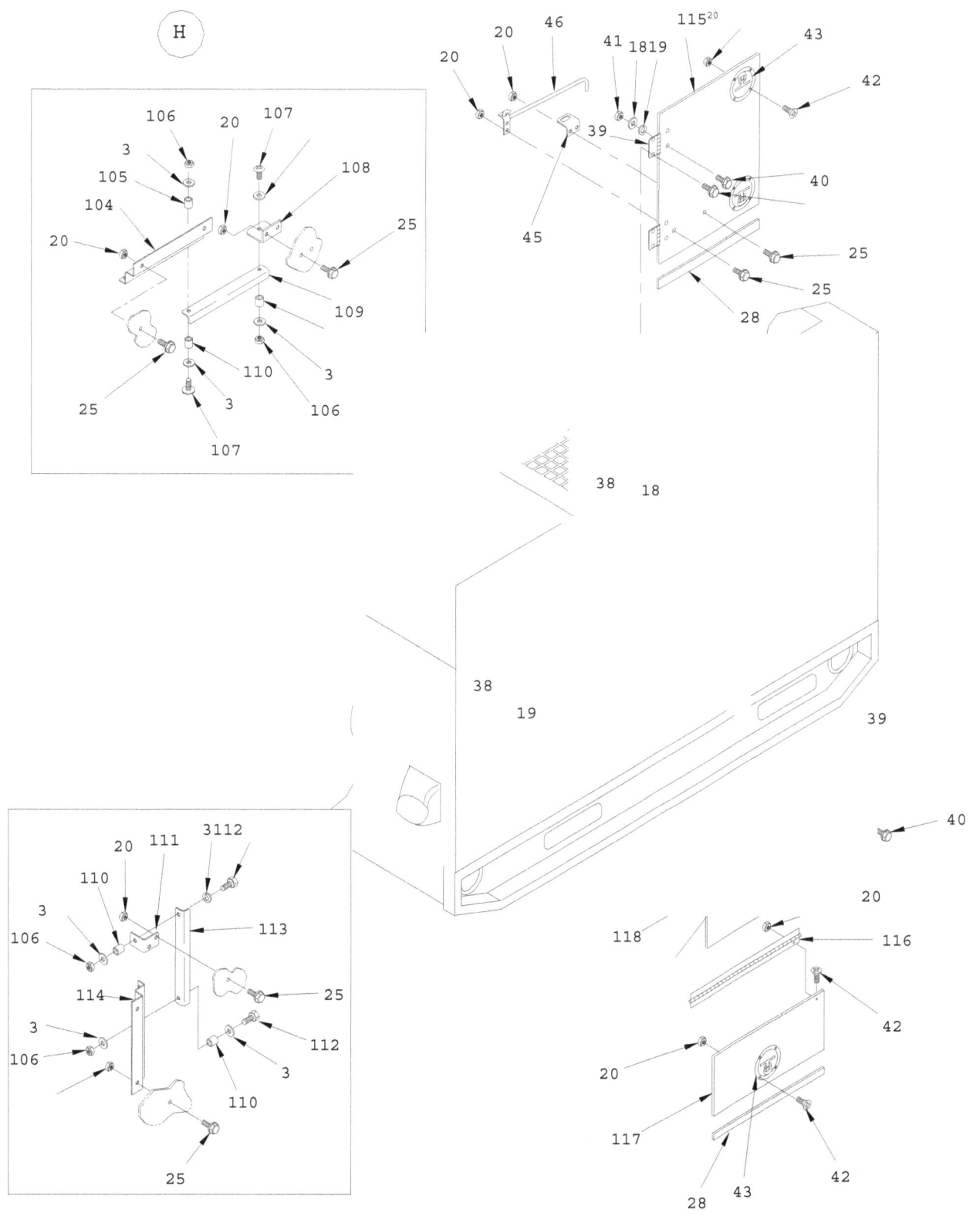

Figure 4. Housing (Sheet 9 of 10)

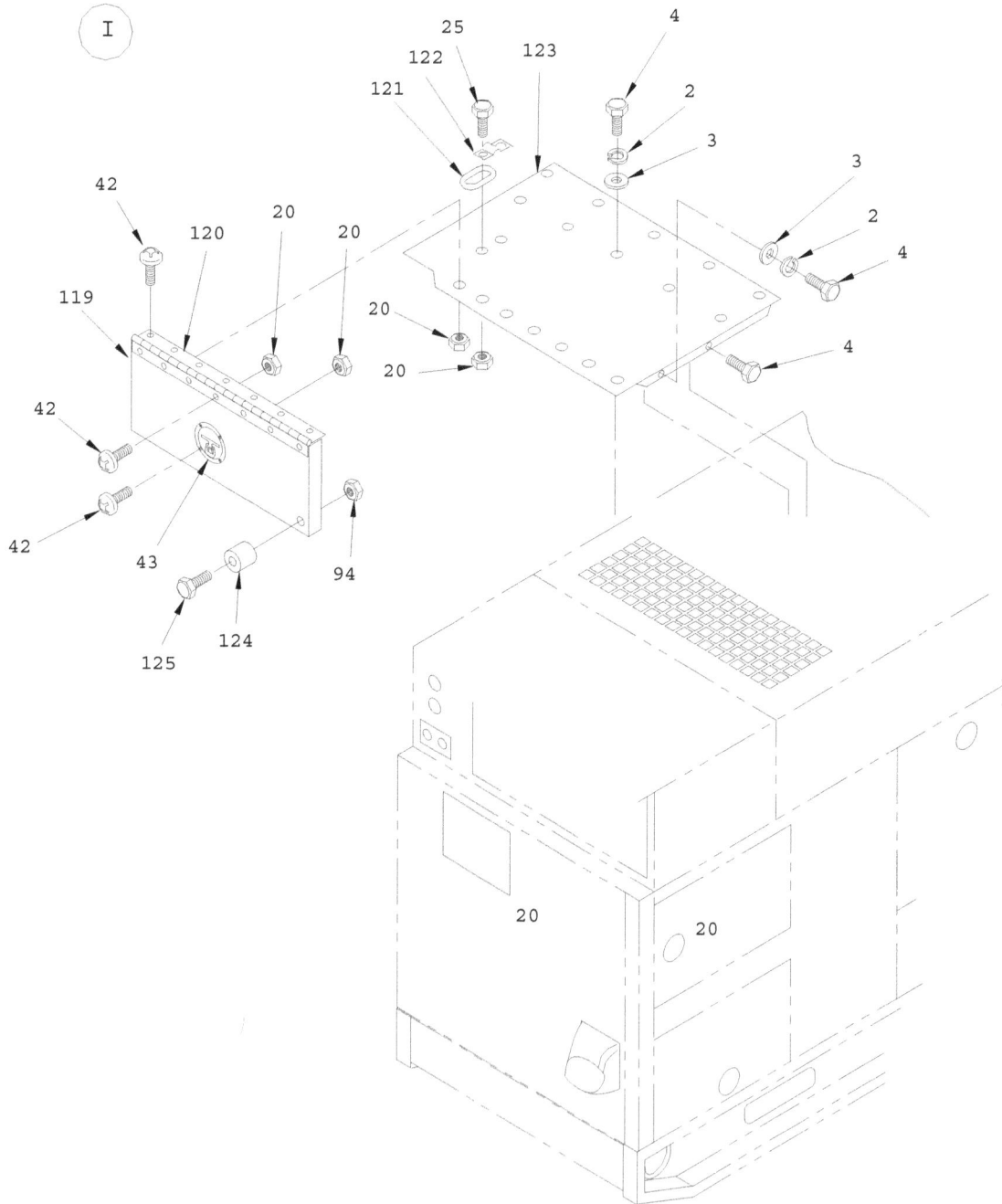

Figure 4. Housing (Sheet 10 of 10)

(1)	(2)			(3)	(4)	(5)	(6)
					SMR CODE		
ITEM	AIR				PART	DESCRIPTION AND	
NO	ARMY	FORCE	USMC	CAGEC	NUMBER	USABLE ON CODE (UOC)	QTY

GROUP 02 HOUSING
FIGURE 4 HOUSING

1	PAOZZ	PA000	PAOZZ	30554	88-20260-2	SCREW,CAP,HEXAGON H...................58
2	PAOZZ	PA000	PAOZZ	97403	13230E6744-62	WASHER,LOCK-SPRING...................116
3	PAOZZ	PA000	PAOZZ	96906	MS51412-2	WASHER,FLAT.........................128
4	PAOZZ	PA000	PAOZZ	30554	88-20260-23	SCREW,CAP,HEXAGON H...................66
5	PAOZZ	PA000	PAOZZ	78553	C7931-1032-38	NUT,PLAIN,CLINCH.....................55
6	PAOZZ	PA000	PAOZZ	30554	88-22429	COVER,EXHAUST..........................1
7	PAOZZ	PA000	PAOZZ	30554	88-22430	BRACKET,MOUNTING.......................1
8	PAOZZ	PA000	PAOZZ	02768	236-170406-04	PIN-RIVET,GROOVED...................350
9	XBOZZ	XB	XBOZZ	30554	88-21594	TOP,HOUSING............................1
10	MOOZZ	M000	M0000	30554	88-22584	INSULATION,TOP,CENT....................4
						MAKE FROM P/N FF40JM02 (28818)
11	MOOZZ	M000	M0000	30554	88-22585	INSULATION,TOP,REAR....................2
						MAKE FROM P/N FF40JM02 (28818)
12	MOOZZ	M000	M0000	30554	88-22588	INSULATION,TOP,CENT....................1
						MAKE FROM P/N FF40JM02 (28818)
13	MOOZZ	M000	M0000	30554	88-22587	INSULATION,TOP,CENT....................2
						MAKE FROM P/N FF40JM02 (28818)
14	MOOZZ	M000	M0000	30554	88-22583	INSULATION,TOP,FRON....................1
						MAKE FROM P/N FF40JM02 (28818)
15	MOOZZ	M000	M0000	30554	88-22586	INSULATION,TOP,FRON....................1
						MAKE FROM P/N FF40JM02 (28818)
16	XBOZZ	XB	XBOZZ	30554	88-21988	PANEL,FILL,RADIATOR....................1
17	PAOZZ	PA000	PAOZZ	30554	88-20260-30	SCREW,CAP,HEXAGON H....................4
18	PAOZZ	PA000	PAOZZ	97403	13230E6744-63	WASHER,FLAT...........................83
19	PAOZZ	PA000	PAOZZ	96906	MS27183-10	WASHER,FLAT...........................48
20	PAOZZ	PA000	PAOZZ	19617	HW310	NUT,PLAIN,ASSEMBLED..................186
21	MOOZZ	M000	M0000	30554	TYPE II GRADE A	RUBBER,CELLULAR........................1
						PER MIL-R-6130, CUT TO LENGTH
22	XBOZZ	XB	XBOZZ	30554	96-23527	PANEL,FLOOR DUCT.......................1
23	XBOZZ	XB	XBOZZ	30554	88-22723	SUPPORT,SEAL,RADIAT....................1
24	XBOZZ	XB	XBOZZ	30554	88-22092	STIFFENER ASSEMBLY.....................1
25	PAOZZ	PA000	PAOZZ	19207	12325869	BOLT,MACHINE.........................108
26	MOOZZ	M000	M0000	30554	88-22582	INSULATION,PANEL,TO....................2
						MAKE FROM P/N FF40JM02 (28818)
27	XBOZZ	XB	XBOZZ	30554	88-21987	PANEL,TOP,RS...........................1
28	MOOZZ	M000	M0000	30554	88-22705	GASKET.................................V
						MAKE FROM P/N 20941 (56329), AS REQ
29	XBOZZ	XB	XBOZZ	30554	96-23617	CHANNEL,AIR DUCT.......................2
30	XBOZZ	XB	XBOZZ	30554	96-23524	FLOOR,DUCT,EXHAUST.....................1
31	MOOZZ	M000	M0000	30554	88-22600	INSULATION,DUCT FLO....................1
						MAKE FROM P/N FF40JM02 (28818), AS REQ
32	XBOZZ	XB	XBOZZ	30554	96-23521	PANEL,DUCT,FLOOR.......................1

(1) ITEM NO	(2) ARMY	(2) AIR FORCE	(2) USMC	(3) CAGEC	(4) SMR CODE PART NUMBER	(5) DESCRIPTION AND USABLE ON CODE (UOC)	(6) QTY
33	XB0ZZ	XB	XB0ZZ	30554	96-23525	SUPPORT,STRUCTURAL.....................1	
34	XB0ZZ	XB	XB0ZZ	30554	88-22702	ANGLE,SUPPORT.........................1	
35	M00ZZ	M000	M0000	30554	88-22599	INSULATION,DUCT FLO...................1	
						MAKE FROM P/N FF40JM02 (28818), AS REQ	
36	PA0ZZ	PA000	PA0ZZ	30554	88-22791-1	SCREW,MACHINE.........................1	
37	XB0ZZ	XB	XB0ZZ	30554	88-21595	PANEL,TOP,LS..........................1	
38	PA0ZZ	PA000	PA0ZZ	96906	MS21266-2N	GROMMET,PLASTIC EDG...................1	
39	XB0ZZ	XB	XB0ZZ	30554	88-21098	HINGE,BUTT...........................12	
40	PA0ZZ	PA000	PA0ZZ	30554	88-20260-31	BOLT,MACHINE.........................25	
41	PA0ZZ	PA000	PA0ZZ	30554	88-22790-1	NUT 58	
42	PA0ZZ	PA000	PA0ZZ	30554	88-22793-4	SCREW,MACHINE........................58	
43	PA0ZZ	PA000	PA0ZZ	30554	88-21099	LEVER,MANUAL CONTRO...................9	
44	XB0ZZ	XB	XB0ZZ	30554	88-21926	DOOR,REAR,LS..........................1	
45	XB0ZZ	XB	XB0ZZ	30554	88-22500	BRACKET,LATCH HOLD....................4	
46	XB0ZZ	XB	XB0ZZ	30554	88-22472	ROD,DOOR HOLDING......................2	
47	XB0ZZ	XB	XB0ZZ	30554	88-22501	BRACKET,HOLDING ROD...................4	
48	XB0ZZ	XB	XB0ZZ	30554	88-20014	HINGE,DOCUMENT BOX....................1	
49	XB0ZZ	XB	XB0ZZ	30554	88-21998	BOX,DOCUMENT..........................1	
50	PA0ZZ	PA000	PA0ZZ	30554	88-20260-20	BOLT,MACHINE.........................14	
51	XB0ZZ	XB	XB0ZZ	30554	88-20013	LID,DOCUMENT BOX......................1	
52	XB0ZZ	XB	XB0ZZ	30554	88-22073	PANEL,REAR,LS.........................1	
53	M00ZZ	M000	M0000	30554	88-22592	INSULATION,BAFFLE.....................1	
						MAKE FROM P/N FF40JM02 (28818)	
54	XB0ZZ	XB	XB0ZZ	30554	88-22050	BAFFLE,AIR INTAKE.....................1	
55	XB0ZZ	XB	XB0ZZ	30554	88-21977	PANEL,REAR............................1	
56	PA0ZZ	PA000	PA0ZZ	24617	274825	SCREW,CAP,HEXAGON H..................21	
57	XB0ZZ	XB	XB0ZZ	30544	88-21733	HINGE,DOOR,AIR CLEA...................1	
58	PA0ZZ	PA000	PA0ZZ	94222	85-16-400-16	STUD,SELF-LOCKING.....................1	
59	XB0ZZ	XB	XB0ZZ	30554	88-21732	DOOR,ACCESS,AIR CLE...................1	
60	PA0ZZ	PA000	PA0ZZ	94222	85-35-309-56	RECEPTACLE,TURNLOCK...................1	
61	PA0ZZ	PA000	PA0ZZ	94222	85-34-101-20	WASHER,SPLIT..........................1	
62	XB0ZZ	XB	XB0ZZ	30554	69-583	COVER,ACCESS..........................1	
63	XB0ZZ	XB	XB0ZZ	30554	88-21964	BOX,ACCESS,LOAD TER...................1	
64	XB0ZZ	XB	XB0ZZ	71600	1-3-0	BUSHING,INSULATING....................1	
65	PA0ZZ	PA000	PA0ZZ	30554	88-20218	COVER,ELECTRICAL GE...................1	
66	PA0ZZ	PA000	PA0ZZ	30554	88-20260-33	BOLT,MACHINE.........................30	
67	PA0ZZ	PA000	PA0ZZ	96906	MS51412-5	WASHER,FLAT..........................38	
68	PA0ZZ	PA000	PA0ZZ	78553	C7988-1420-38	NUT,PLAIN,CLINCH.....................27	
69	XB0ZZ	XB	XB0ZZ	30554	88-22058	PANEL,RS..............................1	
70	XB0ZZ	XB	XB0ZZ	30554	88-22040	SILL,OUTPUT BOX.......................1	
71	XB0ZZ	XB	XB0ZZ	30554	88-21770	SILL,DOOR,ACCESS......................1	
72	XB0ZZ	XB	XB0ZZ	30554	88-22037	POST,CORNER,REAR......................1	
73	XB0ZZ	XB	XB0ZZ	30554	88-21963	DELFECTOR,AIR,UPPER...................1	
74	XB0ZZ	XB	XB0ZZ	30554	88-21914	HOUSING,FRONT.........................1	
75	XB0ZZ	XB	XB0ZZ	30554	88-22718	STIFFENER,SIDE,SHRO...................1	

(1)	(2)			(3)	(4)	(5)	(6)
		AIR			SMR CODE PART	DESCRIPTION AND	
ITEM NO	ARMY	FORCE	USMC	CAGEC	NUMBER	USABLE ON CODE (UOC)	QTY
76	PAOZZ	PA000	PAOZZ	02768	8070-25-00	PUSH ON NUT...........................V	
77	MOOZZ	M000	M0000	30554	88-22591	INSULATION,FRONT HO...................1	
						MAKE FROM P/N FF40JM02 (28818)	
78	XB0ZZ	XB	XB0ZZ	30554	88-22724	SUPPORT,SEAL,RADIAT...................2	
79	XB0ZZ	XB	XB0ZZ	30554	88-21681	PANEL,RADIATOR,RS.....................1	
80	XB0ZZ	XB	XB0ZZ	30554	88-21953	PANEL,AIR DEFLECTOR...................1	
81	XB0ZZ	XB	XB0ZZ	30554	88-22730	SUPPORT,SEAL,RADIAT...................2	
82	XB0ZZ	XB	XB0ZZ	30554	88-22084	BAFFLE,FRONT..........................2	
83	XB0ZZ	XB	XB0ZZ	30554	88-21973	BAFFLE,REAR...........................2	
84	MOOZZ	M000	M0000	30554	88-22593	INSULATION,BAFFLE.....................1	
						MAKE FROM P/N FF40JM02 (28818)	
85	XB0ZZ	XB	XB0ZZ	30554	88-21951	DOOR,FRONT,RS.........................1	
86	XB0ZZ	XB	XB0ZZ	30554	88-22473	ROD,DOOR HOLDING......................2	
87	XB0ZZ	XB	XB0ZZ	30554	88-21680	PANEL,RADIATOR,LS.....................1	
88	XB0ZZ	XB	XB0ZZ	30554	88-21588	DOOR,FRONT,LS.........................1	
89	XB0ZZ	XB	XB0ZZ	30554	88-22031	SILL,DOOR,RS..........................1	
90	XB0ZZ	XB	XB0ZZ	30554	88-21889	SUPPORT,GROUND ROD....................2	
91	PAOZZ	PA000	PAOZZ	80204	B1821BA025C225N	SCREW,CAP,HEXAGON H...................2	
92	XB0ZZ	XB	XB0ZZ	30554	88-21589	SILL,DOOR,LS..........................1	
93	PAOZZ	PA000	PAOZZ	58536	A-A-55804-I-A	ROD,GROUND............................3	
94	XB0ZZ	XB	XB0ZZ	94222	97-50-170-11	CATCH,CLAMPING........................1	
95	PAOZZ	PA000	PAOZZ	45722	P-15121-17	SCREW,ASSEMBLED WASH..................4	
96	PAOZZ	PA000	PAOZZ	78189	501P06080000AMB	NUT,PLAIN,ASSEMBLED..................12	
97	XB0ZZ	XB	XB0ZZ	30554	88-21815	HINGE,STORAGE BOX.....................1	
98	XB0ZZ	XB	XB0ZZ	30554	88-21814	LID,STORAGE BOX.......................1	
99	XB0ZZ	XB	XB0ZZ	30554	88-22791-2	SCREW,MACHINE.........................4	
100	XB0ZZ	XB	XB0ZZ	30554	88-20123	PLATE,STRIKER.........................4	
101	XB0ZZ	XB	XB0ZZ	30554	88-21813	BOX,STORAGE...........................1	
102	XB0ZZ	XB	XB0ZZ	30554	88-22666	PLATE,SPACER,HINGE....................1	
103	XB0ZZ	XB	XB0ZZ	30554	88-22063	DOOR,BATTERY COMPAR...................1	
104	XB0ZZ	XB	XB0ZZ	30554	88-20461	LINK,DOOR SUPPORT.....................1	
105	PAOZZ	PA000	PAOZZ	30554	88-22483	SPACER,SLEEVE.........................2	
106	PAOZZ	PA000	PAOZZ	08928	21NTE102	NUT,SELF-LOCKING......................6	
107	PAOZZ	PA000	PAOZZ	30554	88-22793-8	SCREW,MACHINE.........................2	
108	XB0ZZ	XB	XB0ZZ	30554	88-22479	SUPPORT,DOOR LINK.....................2	
109	XB0ZZ	XB	XB0ZZ	30554	88-22478	BRACKET,DOOR LINK.....................1	
110	PAOZZ	PA000	PAOZZ	30554	88-22482	SPACER,RING...........................4	
111	XB0ZZ	XB	XB0ZZ	30554	88-22475	ANGLE,DOOR SUPPORT....................1	
112	PAOZZ	PA000	PAOZZ	30554	88-22793-7	SCREW,MACHINE.........................2	
113	XB0ZZ	XB	XB0ZZ	30554	88-22476	LINK,DOOR SUPPORT.....................1	
114	XB0ZZ	XB	XB0ZZ	30554	88-22474	BRACKET,DOOR SUPPOR...................1	
115	XB0ZZ	XB	XB0ZZ	30554	88-21927	DOOR,RS...............................1	
116	XB0ZZ	XB	XB0ZZ	30554	88-21820	HINGE,DOOR,ACCESS.....................1	
117	XB0ZZ	XB	XB0ZZ	30554	88-21771	DOOR,ACCESS,LOAD TE...................1	

(1) ITEM NO	(2) ARMY	 AIR FORCE	 USMC	(3) CAGEC	(4) SMR CODE PART NUMBER	(5) DESCRIPTION AND USABLE ON CODE (UOC)	(6) QTY
118	XBOZZ	XB	XBOZZ	30554	88-21585	DOOR,OUTPUT BOX 50/60 HZ..............1 UOC: LTY	
118	XBOZZ	XB	XBOZZ	30554	88-28815	DOOR,OUTPUT BOX 400 HZ................1 UOC: LTX	
119	XBOZZ	XB	XBOZZ	30554	88-21875	DOOR,CONTROL BOX......................1	
120	XBOZZ	XB	XBOZZ	30554	88-21877	HINGE,DOOR,CONTROL....................1	
121	XBOZZ	XB	XBOZZ	1E045	CAT370	RING,DOOR.............................1	
122	XBOZZ	XB	XBOZZ	1E045	CAT380	RETAINER,DOOR RING....................1	
123	XBOZZ	XB	XBOZZ	30554	88-21870	TOP,CONTROL BOX.......................1	
124	PAOZZ	PA000	PAOZZ	31827	MF6T-TPR87-B	COVER,ACCESS..........................2	
125	PAOZZ	PA000	PAOZZ	45722	P-15121-20	SCREW,ASSEMBLED WAS...................2	

END OF FIGURE

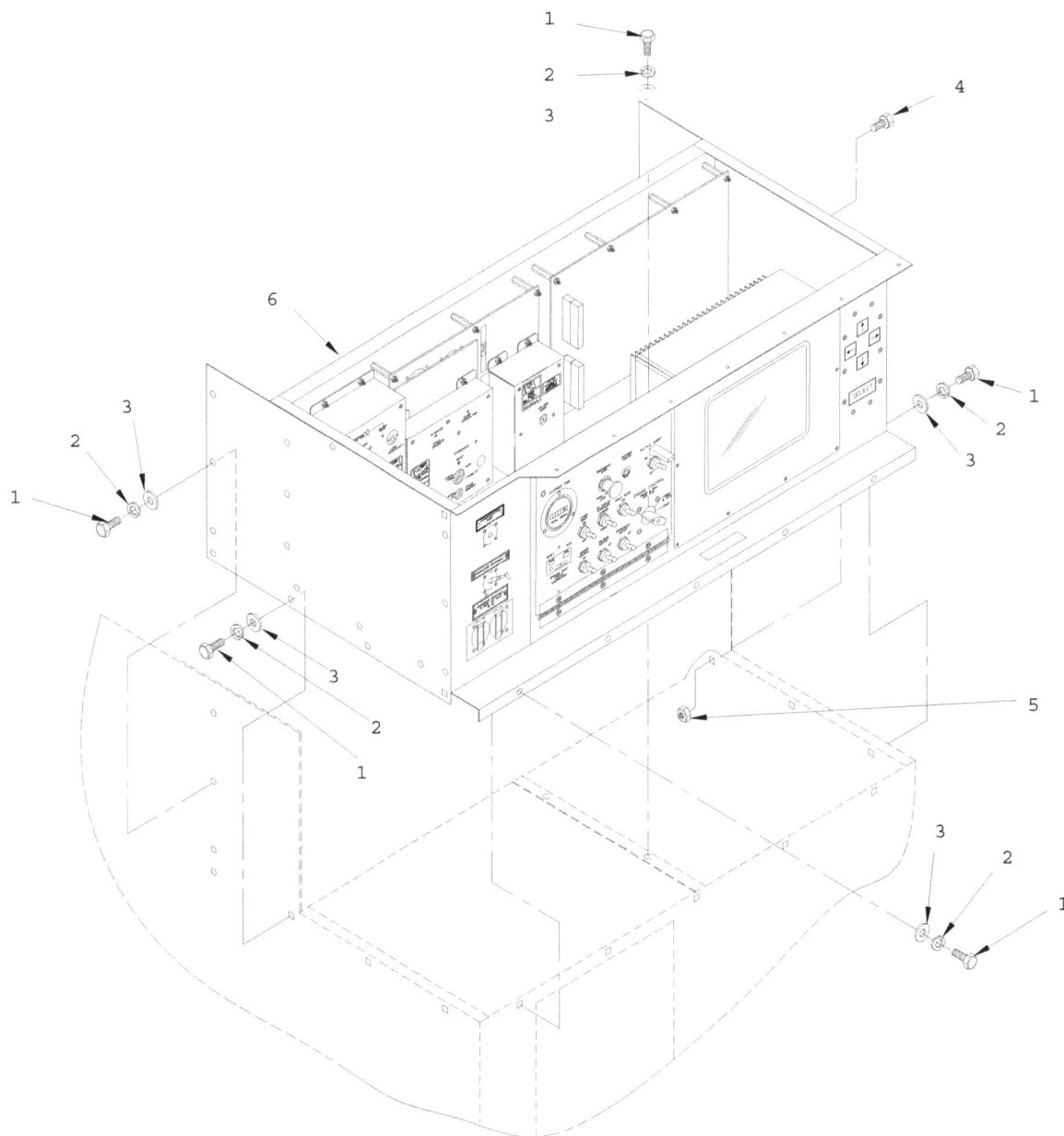

Figure 5. DCS Control Box Assembly (Sheet 1 of 11)

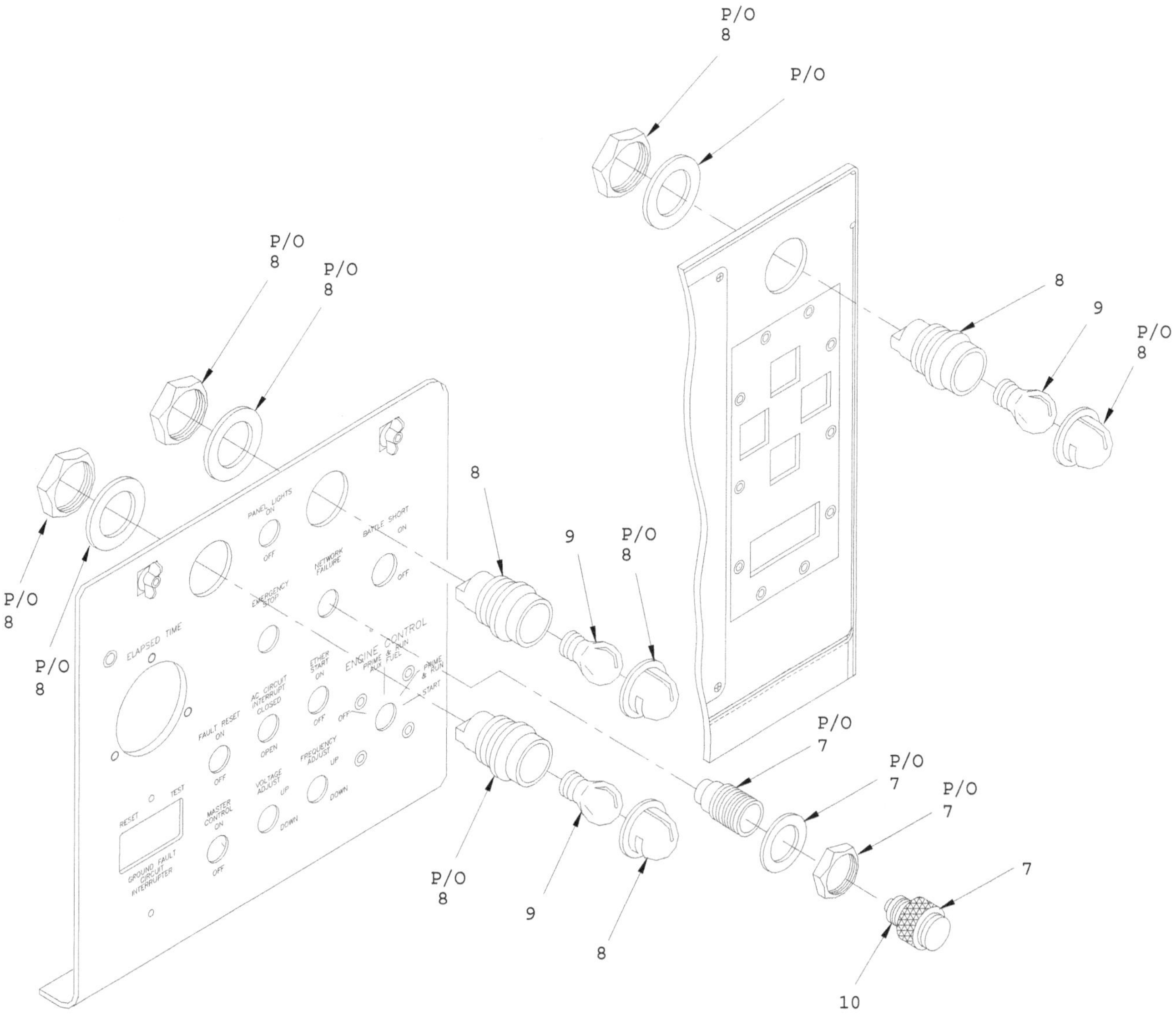

Figure 5. DCS Control Box Assembly (Sheet 2 of 11)

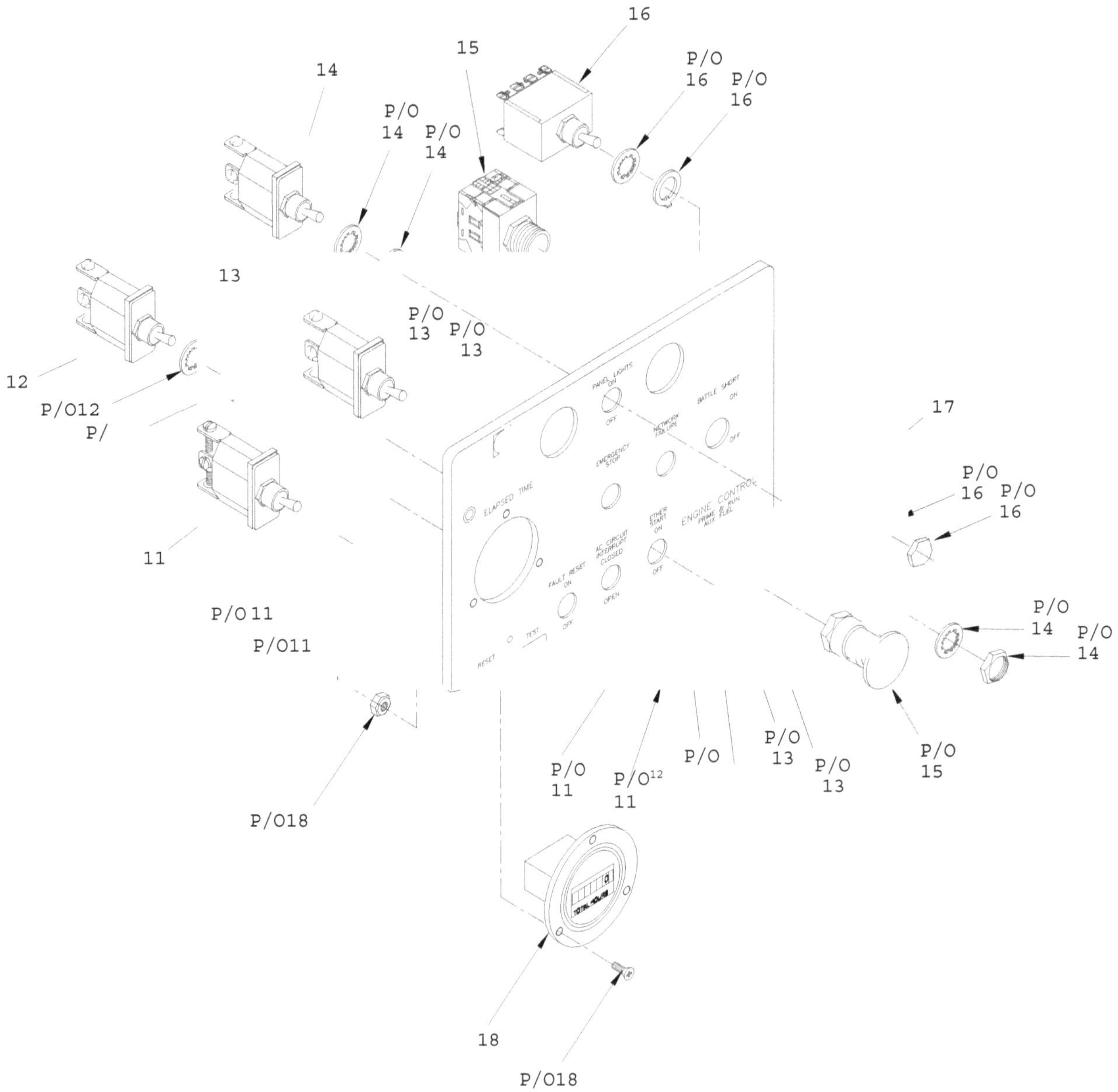

Figure 5. DCS Control Box Assembly (Sheet 3 of 11)

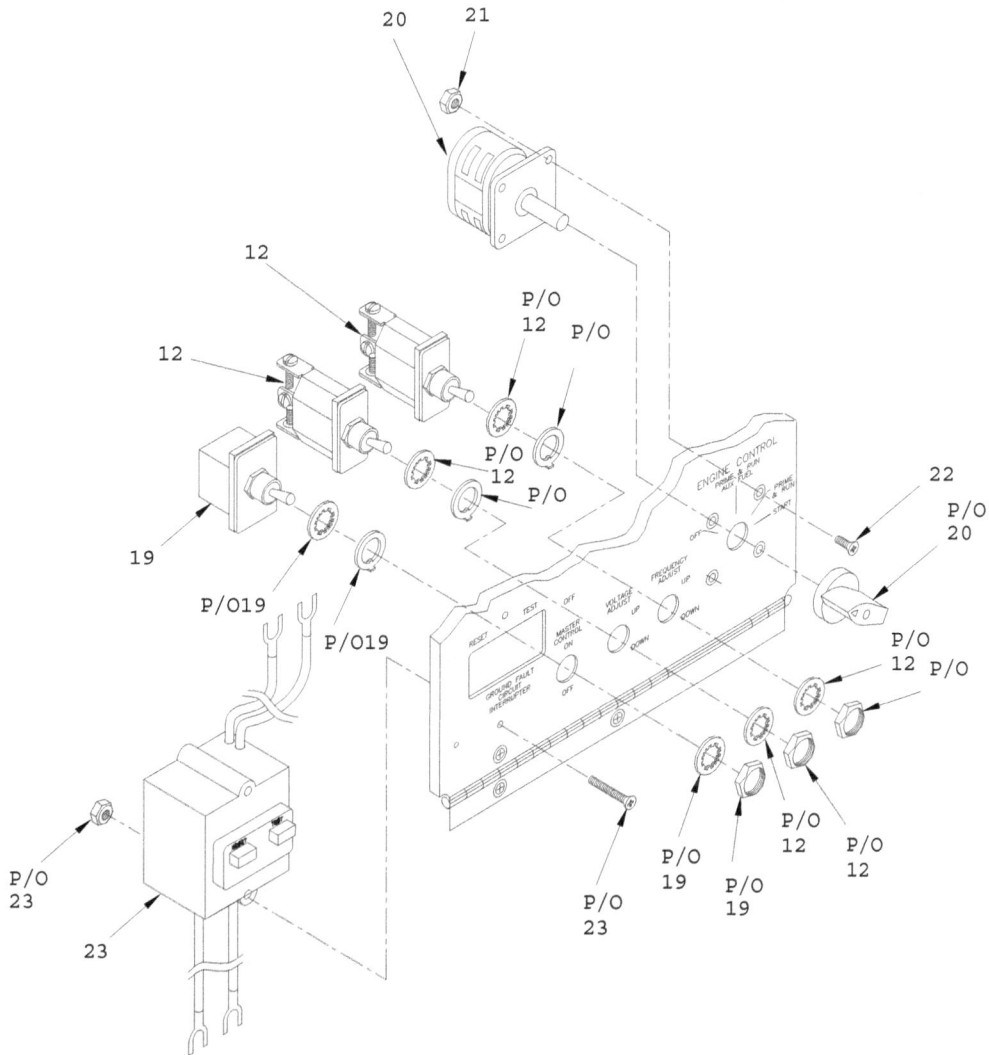

Figure 5. DCS Control Box Assembly (Sheet 4 of 11)

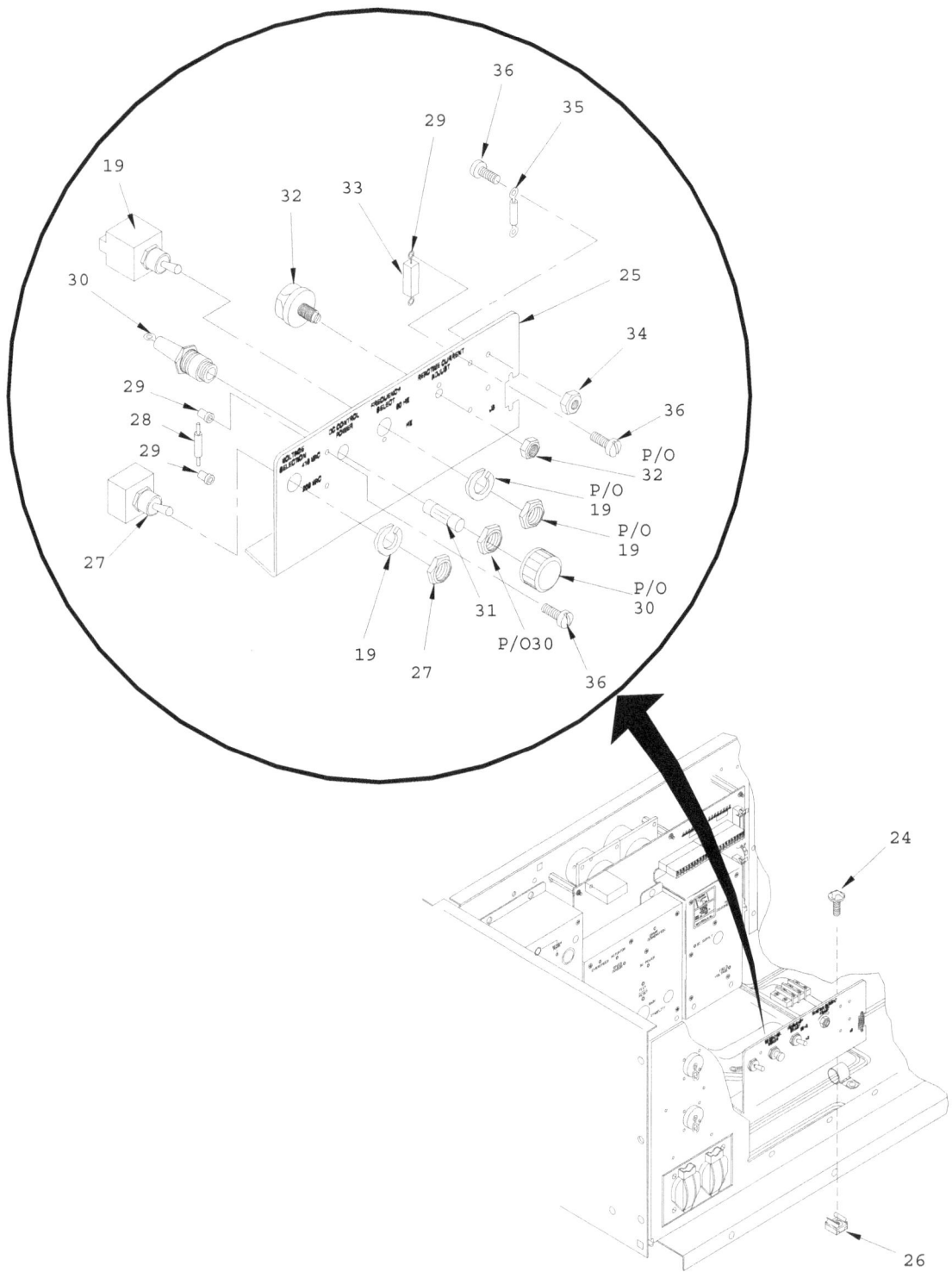

Figure 5. DCS Control Box Assembly (Sheet 5 of 11)

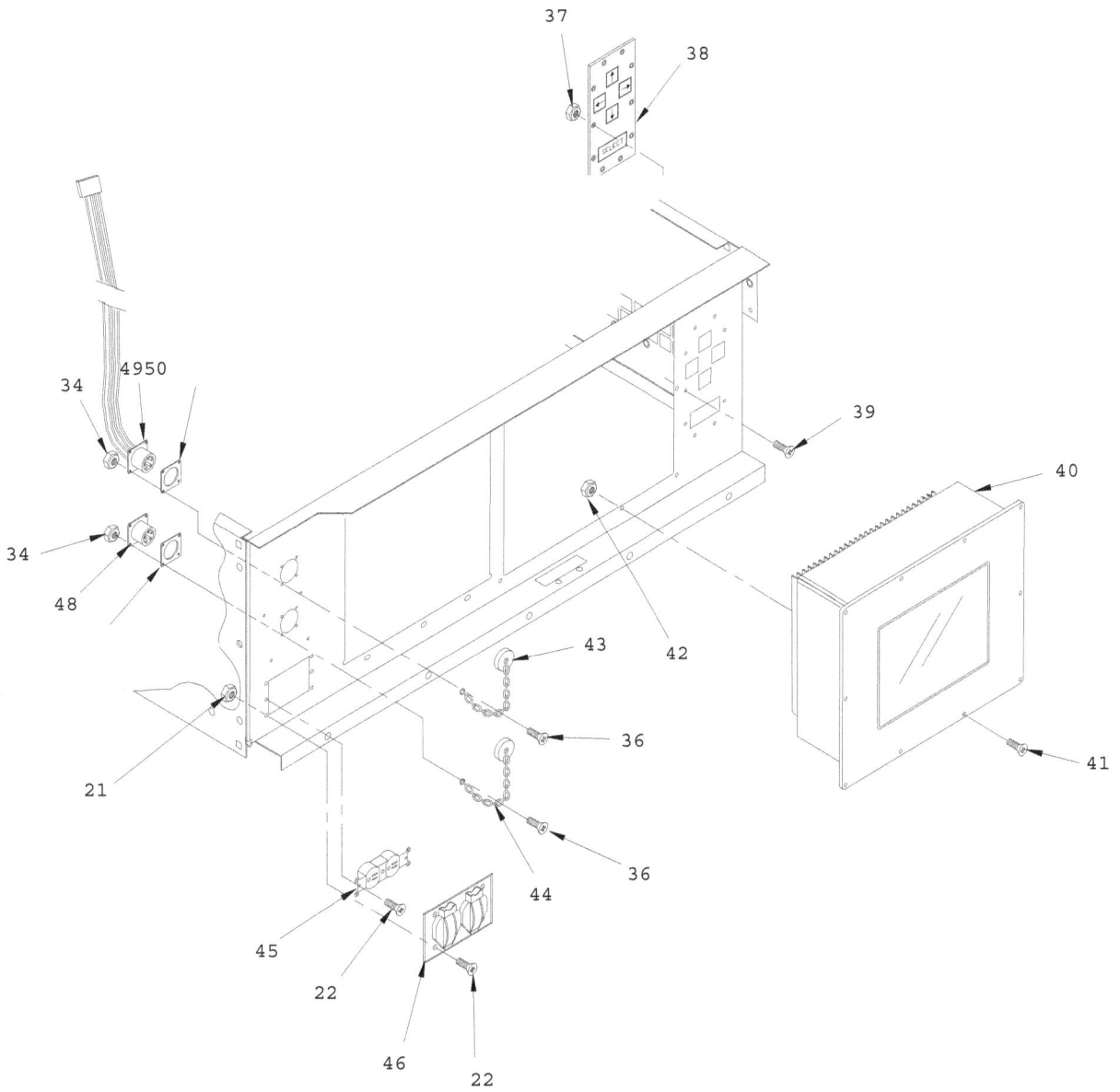

Figure 5. DCS Control Box Assembly (Sheet 6 of 11)

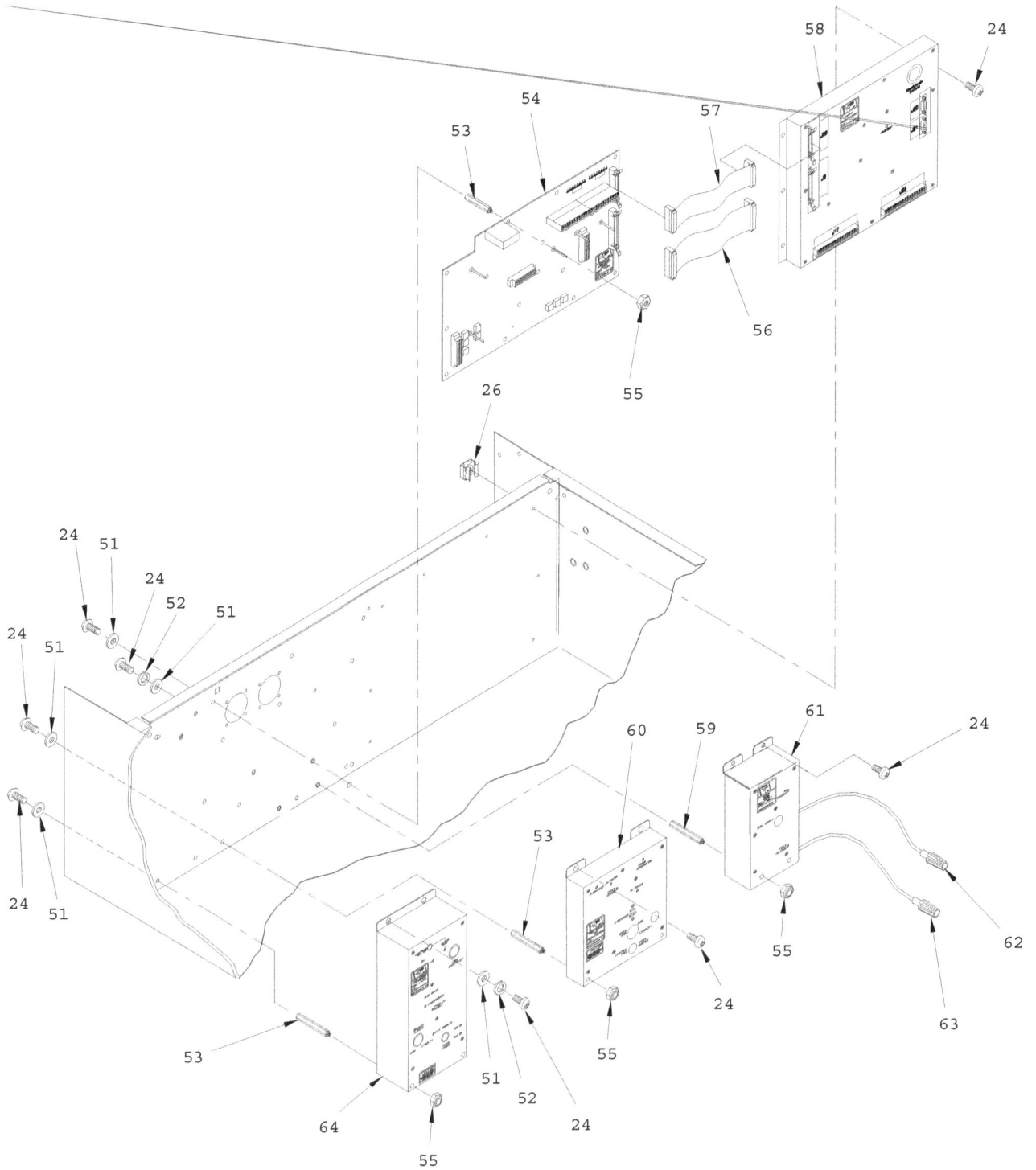

Figure 5. DCS Control Box Assembly (Sheet 7 of 11)

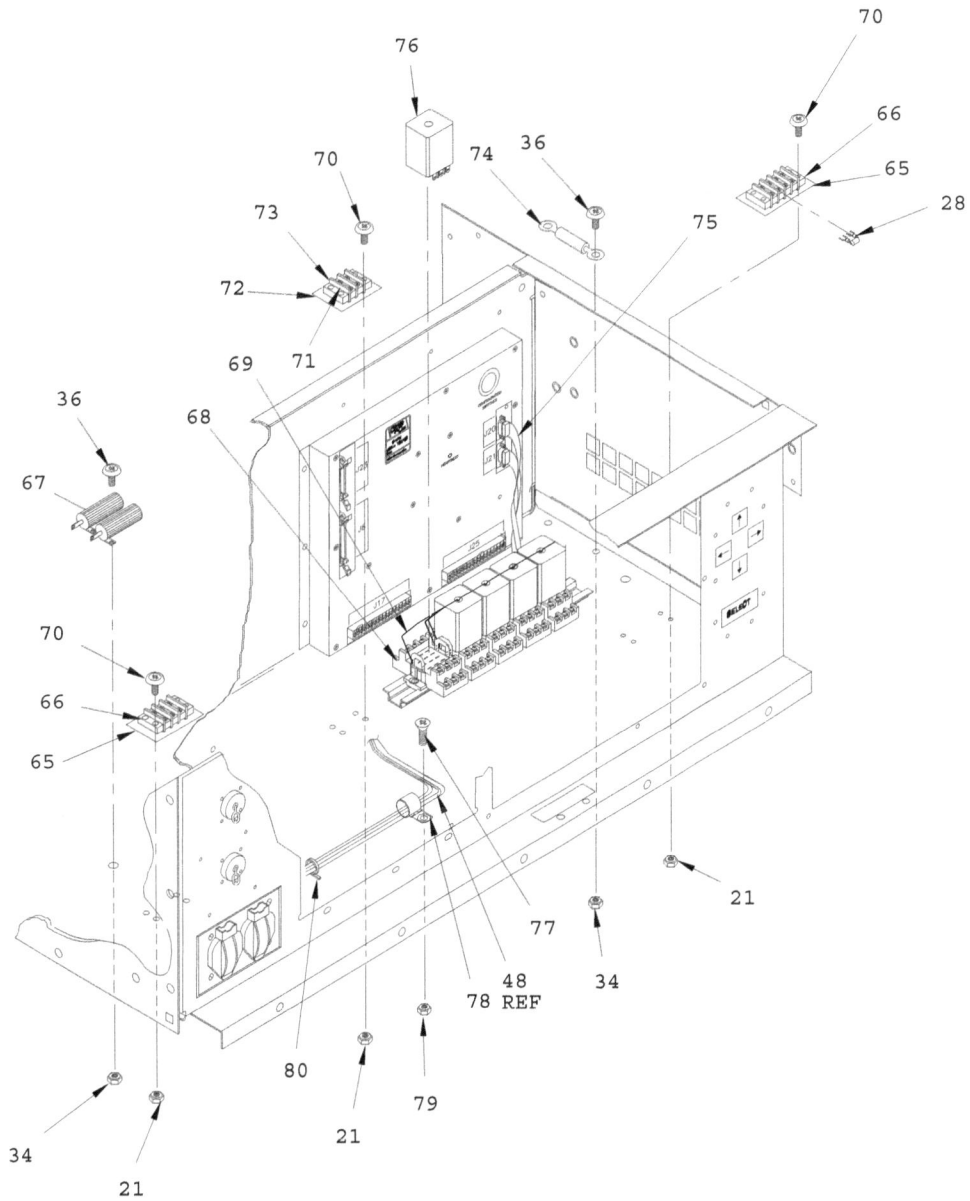

Figure 5. DCS Control Box Assembly (Sheet 8 of 11)

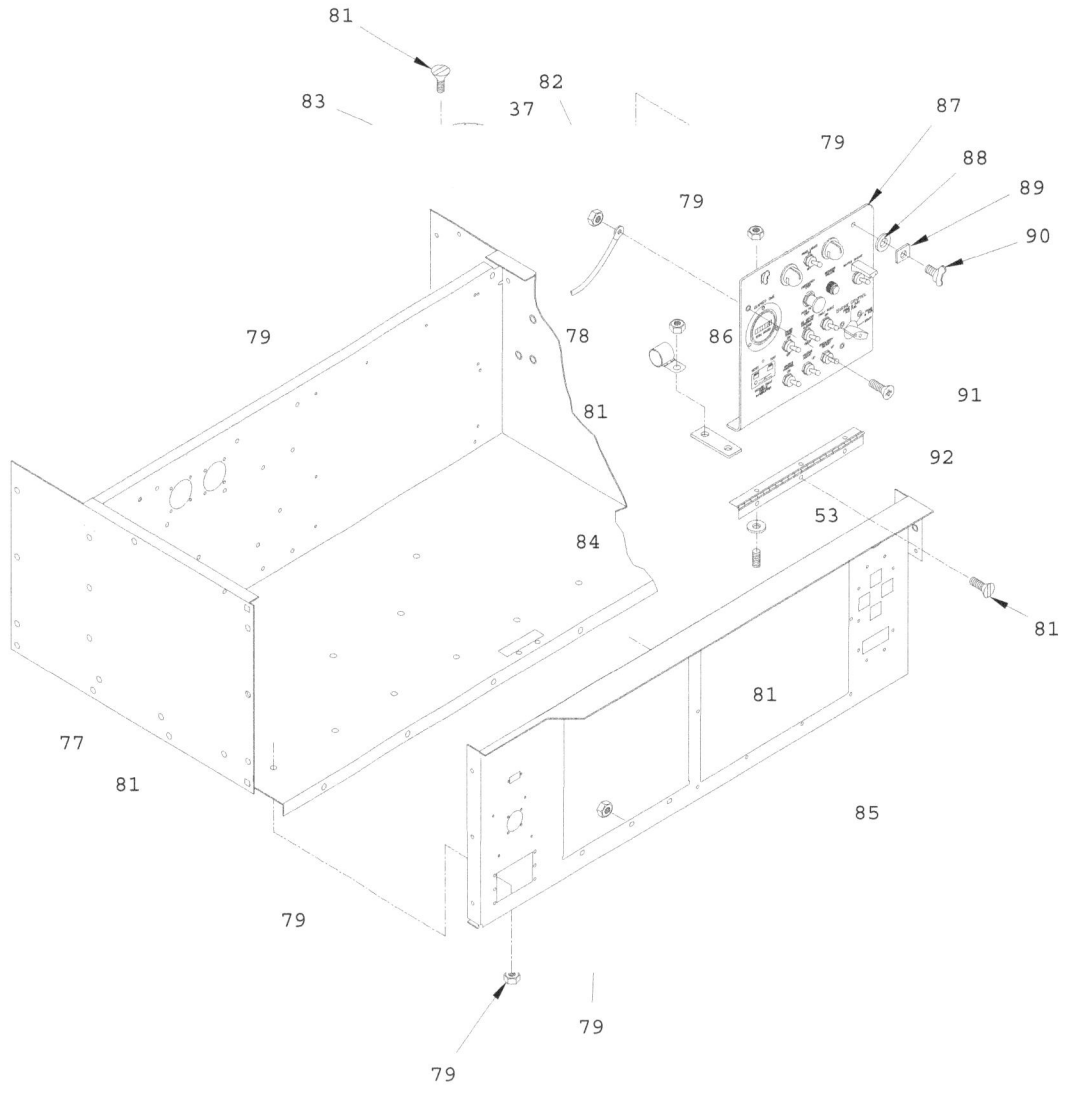

Figure 5. DCS Control Box Assembly (Sheet 9 of 11)

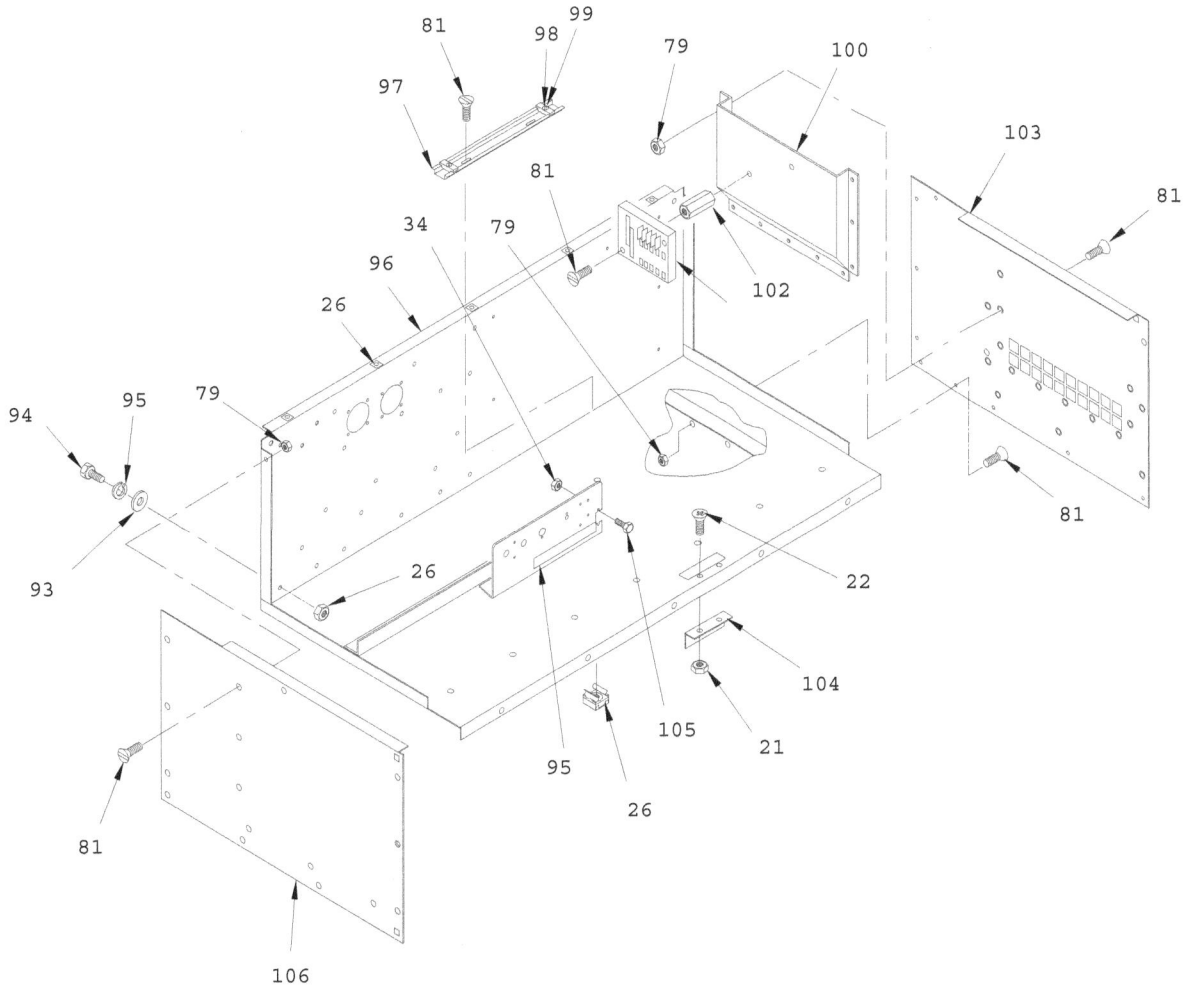

Figure 5. DCS Control Box Assembly (Sheet 10 of 11)

1

107

3

2

5

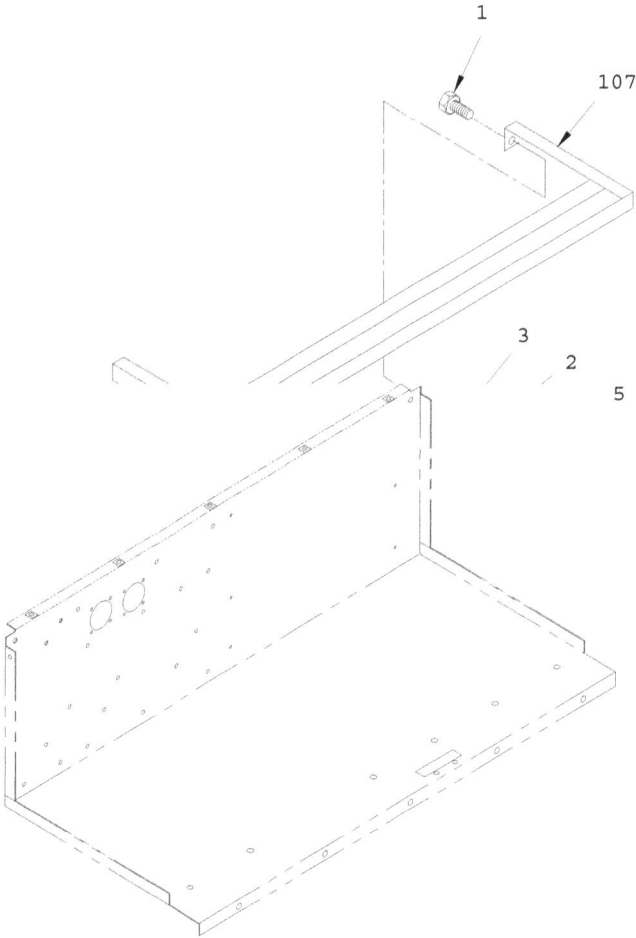

Figure 5. DCS Control Box Assembly (Sheet 11 of 11)

(1)				(3)	(4)	(5)	(6)
		(2)			SMR CODE		
ITEM		AIR			PART	DESCRIPTION AND	
NO	ARMY	FORCE	USMC	CAGEC	NUMBER	USABLE ON CODE (UOC)	QTY

GROUP 03 DCS CONTROL BOX ASSEMBLY
FIGURE 5 DCS CONTROL BOX ASSEMBLY

Item	Army	Air Force	USMC	CAGEC	Part Number	Description	Qty
1	PAOZZ	PA000	PAOZZ	30554	88-20260-23	SCREW,CAP,HEXAGON H.................21
2	PAOZZ	PA000	PAOZZ	97403	13230E6744-62	WASHER,LOCK-SPRING.................21
3	PAOZZ	PA000	PAOZZ	96906	MS51412-2	WASHER,FLAT.................21
4	PAOZZ	PA000	PAOZZ	19207	12325869	BOLT,MACHIN.................5
5	PAOZZ	PA000	PAOZZ	19617	HW310	NUT,PLAIN,ASSEMBLED.................7
6	PB000	PB000	PB000	30554	96-23504	CONTROL BOX,GENERATOR,50/60 HZ........1
						UOC: LTY
6	PB000	PB000	PB000	30554	96-23505	CONTROL BOX,GENERATOR,400 HZ..........1
						UOC: LTX
7	PAOZZ	PA000	PAOZZ	83330	800-1030-0337-50	LIGHT,INDICATOR.....................1
8	PAOZZ	PA000	PAOZZ	71744	6S6/30V-801	LAMP,INCANDESCENT.................3
9	PAOZZ	PA000	PAOZZ	83330	47-0901-2900-201	LIGHT,PANEL.................3
10	PAOZZ	PA000	PAOZZ	1EG71	521-9181	LIGHT EMITTING DIOD.................1
11	PAOZZ	PA000	PAOZZ	81640	8906K4523	SWITCH,TOGGLE.................1
12	PAOZZ	PA000	PAOZZ	19207	12412364	SWITCH,TOGGLE.................3
13	PAOZZ	PA000	PAOZZ	81640	8906K4523	SWITCH,TOGGLE.................1
14	PAOZZ	PA000	PAOZZ	96906	MS24524-22	SWITCH,TOGGLE.................1
15	PAOZZ	PA000	PAOZZ	60886	AYW401-R	SWITCH,PUSH-PULL.................1
16	PAOZZ	PA000	PAOZZ	91929	2TL1-3	SWITCH,TOGGLE.................1
17	PAOZZ	PA000	PAOZZ	96906	MS25224-1	GUARD,SWITCH.................1
18	PAOZZ	PA000	PAOZZ	13555	H400	METER,TIME TOTALIZI.................1
19	PAOZZ	PA000	PAOZZ	81640	8906K4750	SWITCH,TOGGLE.................2
20	PAOZZ	PA000	PAOZZ	10983	C10-C57410E	SWITCH,ROTARY.................1
21	PAOZZ	PA000	PAOZZ	78189	501P06080000AMB	NUT,PLAIN,ASSEMBLED.................19
22	PAOZZ	PA000	PAOZZ	30554	88-22791-2	SCREW,MACHINE.................13
23	PAOZZ	PA000	PAOZZ	60177	29400-2	INTERRUPTER,GROUND.................1
24	PAOZZ	PA000	PAOZZ	30554	88-22793-4	SCREW,MACHINE.................32
25	XBOZZ	XB	XBOZZ	30554	96-23517	BRACKET,MOUNTING.................1
26	PAOZZ	PA000	PAOZZ	78553	C7941-1032-38	NUT,PLAIN,CLINCH.................15
27	PAOZZ	PA000	PAOZZ	81640	8530K9	SWITCH,TOGGLE.................1
28	PAOZZ	PA000	PAOZZ	30554	88-22418-1	SEMICONDUCTOR DEVIC.................8
29	PAOZZ	PA000	PAOZZ	18310	1127-38-0516	TERMINAL,STUD.................4
30	PAOZZ	PA000	PAOZZ	81349	FHN26G1	FUSEHOLDER,EXTRACTO.................1
31	PAOZZ	PA000	PAOZZ	81495	1223-299	FUSE,CARTRIDGE.................1
32	PAOZZ	PA000	PAOZZ	44655	REL5R0	RESISTOR,VARIABLE,W.................1
33	PAOZZ	PA000	PAOZZ	7T184	5W133	RESISTOR,FIXED,WIRE.................1
34	PAOZZ	PA000	PAOZZ	78189	501-040800-00	NUT,PLAIN,ASSEMBLED.................18
35	PAOZZ	PA000	PAOZZ	91637	HLM-10-10Z130IJ	RESISTOR,FIXED,WIRE.................1
36	PAOZZ	PA000	PAOZZ	30554	69-662-5	SCREW,ASSEMBLED WAS.................18
37	PAOZZ	PA000	PAOZZ	78189	511-081800-00	NUT,PLAIN,ASSEMBLED.................11
38	PAOZZ	PA000	PAOZZ	0FNW8	96-23545	KEYPAD ASSEMBLY.................1

(1)	(2)		(3)	(4)	(5)	(6)
	AIR			SMR CODE	DESCRIPTION AND	
ITEM				PART		
NO	ARMY	FORCE	USMC	CAGEC	NUMBER	DESCRIPTION AND USABLE ON CODE (UOC)	QTY

ITEM NO	ARMY	AIR FORCE	USMC	CAGEC	PART NUMBER	DESCRIPTION AND USABLE ON CODE (UOC)	QTY
39	PAOZZ	PA000	PAOZZ	30554	88-20260-11	SCREW,ASSEMBLED WAS...................12
40	PAOZZ	PA000	PAOZZ	0FNW8	96-23569	DISPLAY UNIT...........................1
41	PAOZZ	PA000	PAOZZ	45722	P-15121-38	SCREW,ASSEMBLED WAS...................12
42	PAOZZ	PA000	PAOZZ	78276	RNS832-120KT2	NUT,PLAIN,BLIND RIV...................10
43	PAOZZ	PA000	PAOZZ	77820	9760-22	CAP,PROTECTIVE,DUST....................1
44	PAOZZ	PA000	PAOZZ	06324	660-005C18S4.5-0	COVER,ELECTRICAL,CO....................1
45	PAOZZ	PA000	PAOZZ	74545	CR15	CONNECTOR BODY,RECE....................1
46	PAOZZ	PA000	PAOZZ	64533	RP-5	PLATE,WALL,ELECTRIC....................1
47	PAOZZ	PA000	PAOZZ	77820	10-40450-18	GASKET.................................1
48	PA000	PA000	PA000	30554	96-23565	WIRING HARNESS,CONT....................1
						(SEE FIGURE 6 FOR PARTS BREAKDOWN)
49	PAOZZ	PA000	PAOZZ	30554	96-23683	CABLE ASSEMBLY,SPEC....................1
50	PAOZZ	PA000	PAOZZ	77820	10-40450-22	GASKET.................................1
51	PAOZZ	PA000	PAOZZ	96906	MS51412-2	WASHER,FLAT...........................44
52	PAOZZ	PA000	PAOZZ	97403	1320E6744-62	WASHER,LOCK...........................38
53	PAOZZ	PA000	PAOZZ	61000	A7416-1032-0	STANDOFF,PLAIN,SNAP...................24
54	PAOZZ	PA000	PAOZZ	0BXW5	TCM102	PRINTED WIRING BOAR....................1
55	PAOZZ	PA000	PAOZZ	08928	21NTE102	NUT,SELF-LOCKING......................22
56	PAOZZ	PA000	PAOZZ	0BXW5	96-23665	CABLE ASSEMBLY,PRIN....................1
57	PAOZZ	PA000	PAOZZ	0BXW5	96-23664	CABLE ASSEMBLY,PRIN....................1
58	PAOZZ	PA000	PAOZZ	0BXW5	TCM100-50/60HZ	BACKPLANE ASSEMBLY.....................1
						UOC: LTY
58	PAOZZ	PA000	PAOZZ	0BXW5	TCM400-400HZ	BACKPLANE ASSEMBLY.....................1
						UOC: LTX
59	PAOZZ	PA000	PAOZZ	0GCW4	A7436-1032-0	STANDOFF,PLAIN,SNAP....................2
60	PAOZZ	PA000	PAOZZ	0BXW5	ESD5551	MOTOR,GOVERNOR.........................1
61	PAOZZ	PA000	PAOZZ	30554	96-23542	REGULATOR,VOLTAGE......................1
						UOC: LTY
61	PAOZZ	PA000	PAOZZ	0BXW5	AVR100	REGULATOR,VOLTAGE......................1
						UOC: LTX
62	PAOZZ	PA000	PAOZZ	56501	RB25177M	TERMINAL,QUICK DISC....................1
63	PAOZZ	PA000	PAOZZ	56501	RB2573M	TERMINAL,QUICK DISC....................1
64	PAOZZ	PA000	PAOZZ	0BXW5	LSS100	SYNCHRONIZER,ELECTR....................1
						UOC: LTY
64	PAOZZ	PA000	PAOZZ	0BXW5	LSS400	SYNCHRONIZER,ELECTR....................1
						UOC: LTX
65	XBOZZ	XB	XBOZZ	9R803	3300-4-XP-74	MARKER STRIP...........................2
66	PAOZZ	PA000	PAOZZ	9R803	3300-4	TERMINAL BOX...........................2
67	PAOZZ	PA000	PAOZZ	30554	96-23666	RESISTOR,FIXED,WIRE....................2
68	PAOZZ	PA000	PAOZZ	77342	27E893	SOCKET,PLUG-IN ELEC....................5
69	PAOZZ	PA000	PAOZZ	77342	20C318	CLIP,SPRING TENSION....................5
70	PAOZZ	PA000	PAOZZ	30554	P-15121-36	SCREW,ASSEMBLED WAS....................8
71	PAOZZ	PA000	PAOZZ	9R803	330JS	CONTACT,ELECTRICAL.....................1
72	XBOZZ	XB	XBOZZ	9R803	3300-2-XP-74	MARKER STRIP,TERMIN....................1
73	PAOZZ	PA000	PAOZZ	9R803	3300-2	TERMINAL BOARD.........................1

(1)				(3)	(4)	(5)	(6)
					SMR CODE		
ITEM		AIR			PART	DESCRIPTION AND	
NO	ARMY	FORCE	USMC	CAGEC	NUMBER	USABLE ON CODE (UOC)	QTY

		(2)					

74 PAOZZ PAOOO PAOZZ 30554 88-20263 RESISTOR,FIXED,WIRE...................1

75	PAOZZ	PAOOO	PAOZZ	OFNW8	96-23574	WIRING HARNESS,BRAN..................1
76	PAOZZ	PAOOO	PAOZZ	77342	KUP14D15-24VDC	RELAY,ELECTROMAGNET..................5
77	PAOZZ	PAOOO	PAOZZ	30554	88-22793-6	SCREW,MACHINE........................3

78 PAOZZ PAOOO PAOZZ 30554 88-20546-4 CLAMP,LOOP..........................2

| 79 | PAOZZ | PAOOO | PAOZZ | 19617 | HW310 | NUT,PLAIN,ASSEMBLED.................56 | 80 |

PAOZZ PAOOO PAOZZ 96906 MS3367-5-9 STRAP,TIEDOWN,ELECT...................V

81	PAOZZ	PAOOO	PAOZZ	19207	12325869	BOLT,MACHINE........................51
82	XBOZZ	XB	XBOZZ	30554	88-22120	HOLDER,CONTROL PANE..................1
83	XBOZZ	XB	XBOZZ	30554	88-22774	SCREEN,VENT..........................1
84	PAOZZ	PAOOO	PAOZZ	94222	85-35-309-56	RECEPTACLE,TURNLOCK..................2
85	XBOZZ	XB	XBOZZ	30554	96-23516	FRAME,CONTROL PANEL..................1
86	XBOZZ	XB	XBOZZ	30554	96-23673	BRACKET,MOUNTING.....................1
87	XBOZZ	XB	XBOZZ	30554	96-23508	PANEL,CONTROL,ELECT..................1
88	PAOZZ	PAOOO	PAOZZ	94222	85-46-103-39	WASHER,FLAT..........................2
89	PAOZZ	PAOOO	PAOZZ	94222	85-34-101-20	WASHER,SPLIT.........................2
90	PAOZZ	PAOOO	PAOZZ	94222	85-12-460-16	STUD,TURNLOCK FASTE..................2
91	PAOZZ	PAOOO	PAOZZ	30554	88-20555-2	SCREW,MACHINE........................1
92	XBOZZ	XB	XBOZZ	30554	96-23519	HINGE,CONTROL PANEL..................1
93	PAOZZ	PAOOO	PAOZZ	30554	88-20565-2	SCREW,MACHINE........................2
94	PAOZZ	PAOOO	PAOZZ	30554	88-20260-31	BOLT,MACHINE.........................2
95	PAOZZ	PAOOO	PAOZZ	97403	1320E6744-63	WASHER,FLAT..........................2
96	XBOZZ	XB	XBOZZ	30554	96-23518	PANEL,BACK AND BOTT..................1
97	PAOZZ	PAOOO	PAOZZ	30554	88-20258	BRACKET,MOUNTING.....................1
98	PAOZZ	PAOOO	PAOZZ	77342	40G432	INSULATOR 2
99	PAOZZ	PAOOO	PAOZZ	77342	24A071	MOUNTING CLIP........................2
100	XBOZZ	XB	XBOZZ	30554	96-23685	BAFFLE,AIRFLOW.......................1
101	PAOZZ	PAOOO	PAOZZ	0BXW5	PCI102	POWER SUPPLY.........................1
102	PAOZZ	PAOOO	PAOZZ	61000	A3912-832-0	STANDOFF,THREADED,S..................2
103	XBOZZ	XB	XBOZZ	30554	96-23635	PANEL,CONTROL BOX,R..................1
104	XBOZZ	XB	XBOZZ	30554	88-21874	ANGLE,LATCH..........................1

105 PAOZZ PAOOO PAOZZ 30554 96-23616 JACKSCREW,ELECTRICA..................2

| 106 | XBOZZ | XB | XBOZZ | 30554 | 96-23634 | PANEL,CONTROL BOX,L..................1 |
| 107 | XBOZZ | XB | XBOZZ | 30554 | 96-23556 | STIFFENER............................1 |

END OF FIGURE

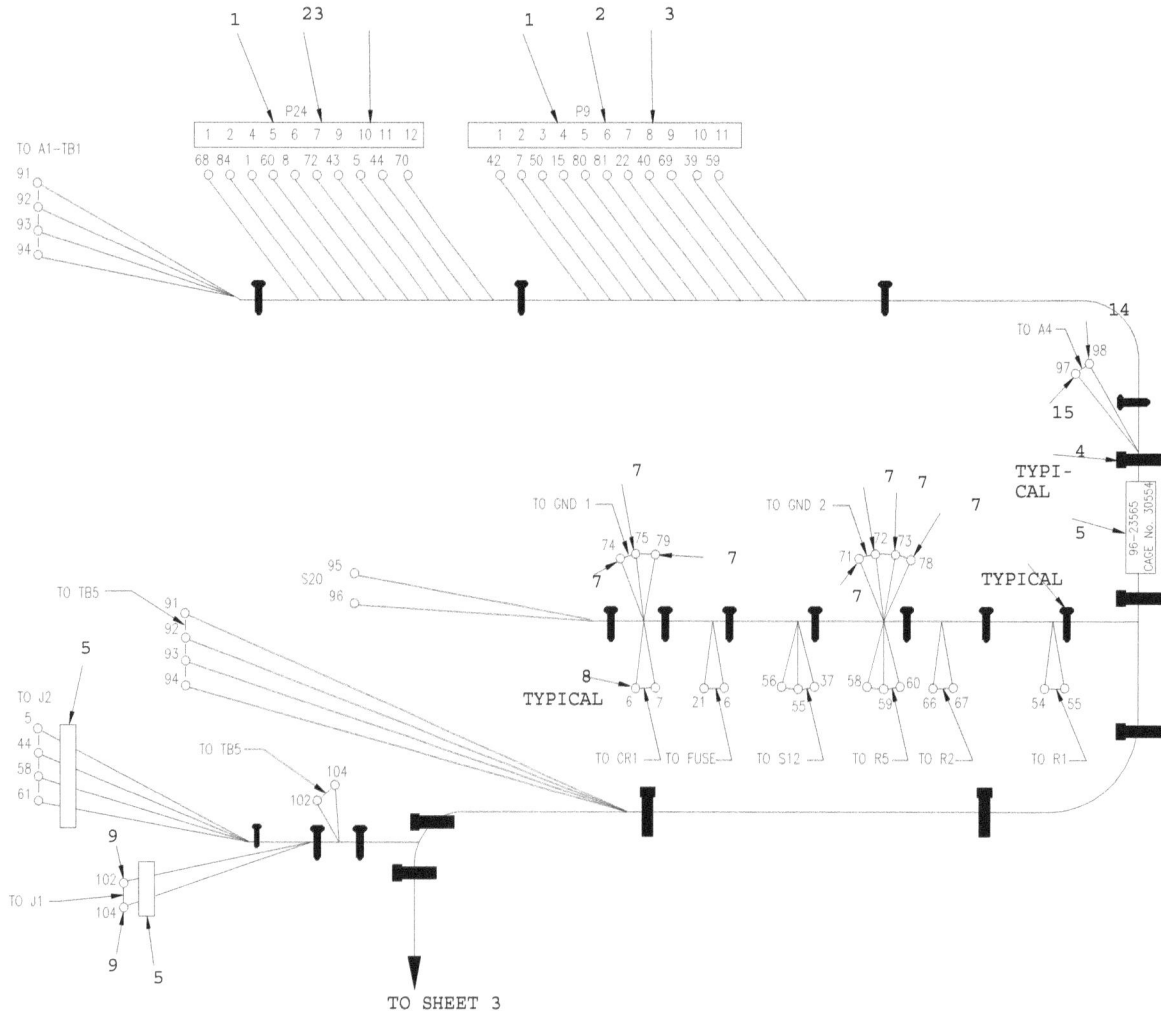

Figure 6. Control Box Wiring Harness (Sheet 1 of 3)

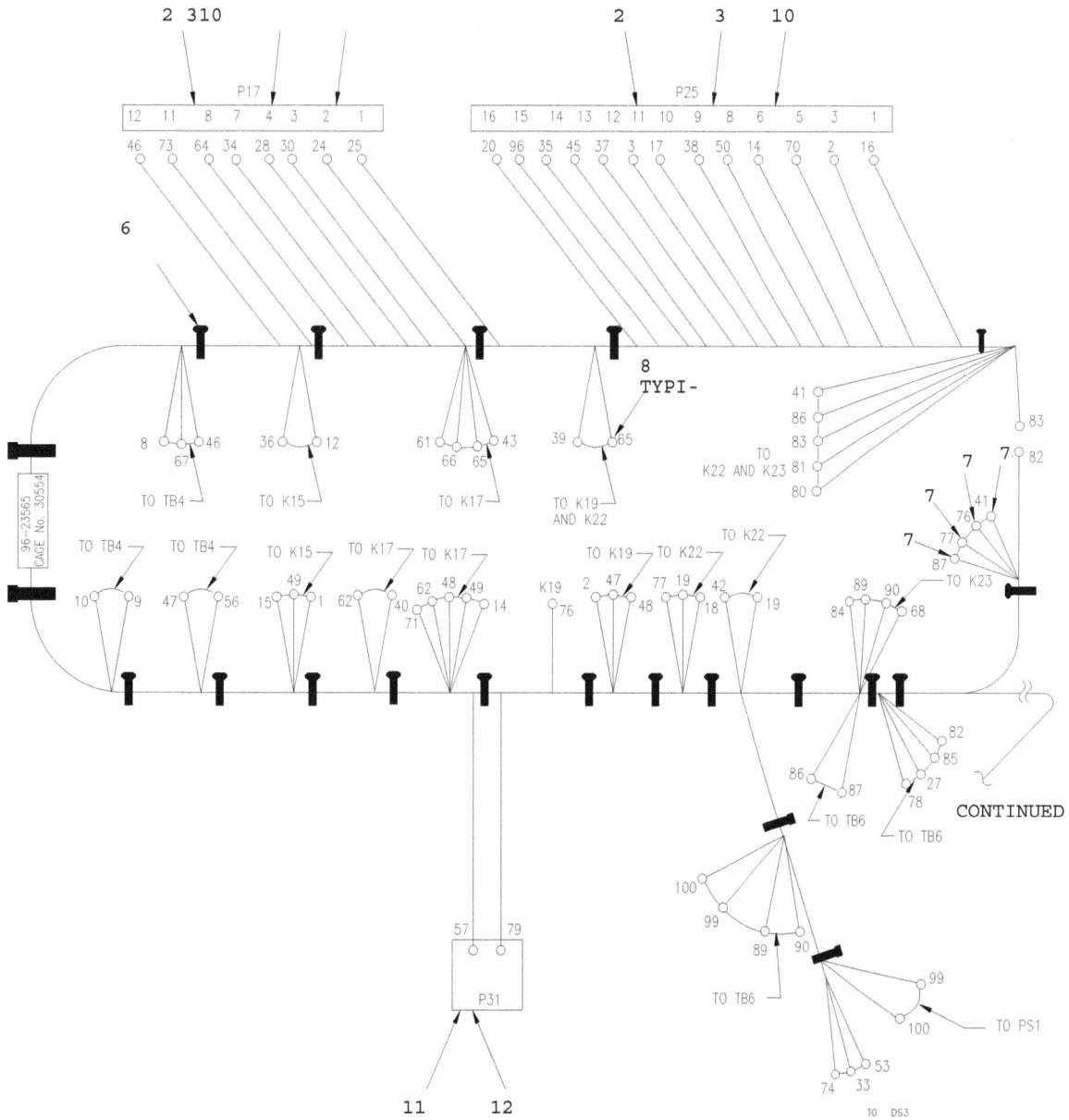

Figure 6. Control Box Wiring Harness (Sheet 2 of 3)

FRO M SHEET 1

8
TYPICAL

TO S18   TO S19   TO S1
TO S3   22 23   29 30 26   24 26   35
         21   28   29   25   88
        57   20

13   6
                                         TYPI-

TO S5   13 17 18   TO S9
to M3   4 +   11   3   12 69   63
                                               45
        51 -   4   TO S1

        34   TO DS5   16
    13   54
13   95 27   75
TO S4

                           38   64
TO S1   TO S17   63   11 TO S7
TO DS2   63   10
TO DS1   32   88
        31   31   52 53   36   TO S7
        52   23   33   85
        51   32   32

15   109
            97
6

6

FRO M SHEET 2   109
                              98
        6   14

R16 AND R17 BRANCH

Figure 6. Control Box Wiring Harness (Sheet 3 of 3)

(1)		(2)		(3)	(4) SMR CODE	(5)	(6)
ITEM NO	ARMY	AIR FORCE	USMC	CAGEC	PART NUMBER	DESCRIPTION AND USABLE ON CODE (UOC)	QTY

GROUP 0321 DCS CONTROL BOX WIRING HARNESS
FIGURE 6 DCS CONTROL BOX WIRING HARNESS

1	PAOZZ	PAO00	PAOZZ	5Y407	1792346	CONNECTOR,PLUG........................2	
2	PAOZZ	PAO00	PAOZZ	5Y407	5600014	MARKER CARD..........................4	
3	PAOZZ	PAO00	PAOZZ	5Y407	5600030	MARKER CARD..........................4	4
PAOZZ	PAO00	PAOZZ	06383	PLT2S		STRAP,TIEDOWN,ELECT....................V	
5	PAOZZ	PAO00	PAOZZ	96906	MS3368-1-9-A	STRAP,IDENTIFICATIO...................8	6
PAOZZ	PAO00	PAOZZ	96906	MS3367-5-9	STRAP,TIEDOWN,ELECT....................V		
7	PAOZZ	PAO00	PAOZZ	98410	BB-837-10	TERMINAL,LUG.........................11	
8	PAOZZ	PAO00	PAOZZ	98410	BB-8707-06	TERMINAL,SPADE......................122	
9	PAOZZ	PAO00	PAOZZ	98410	BB-8194-08	TERMINAL,SPADE........................2	
10	PAOZZ	PAO00	PAOZZ	5Y407	1792391	CONNECTOR,PLUG........................2	
11	PAOZZ	PAO00	PAOZZ	30554	96-23650	PLUG,D-SUBMINIATURE...................1	
12	PAOZZ	PAO00	PAOZZ	30554	96-23651	BACKSHELL,PLUG........................1	
13	PAOZZ	PAO00	PAOZZ	98410	AA-8714-08	TERMINAL,DISCONNECT...................2	
14	PAOZZ	PAO00	PAOZZ	15912	RB2573	TERMINAL,QUICK DISC...................3	
15	PAOZZ	PAO00	PAOZZ	56501	RB25177M	TERMINAL,QUICK DISC...................1	16
	MOOZZ	MO00	MO000	30554	88-20450-4	WIRE,ELECTRICAL.......................V	

MAKE FROM P/N M5086/2-16-9 (81349),

AS REQ

END OF FIGURE

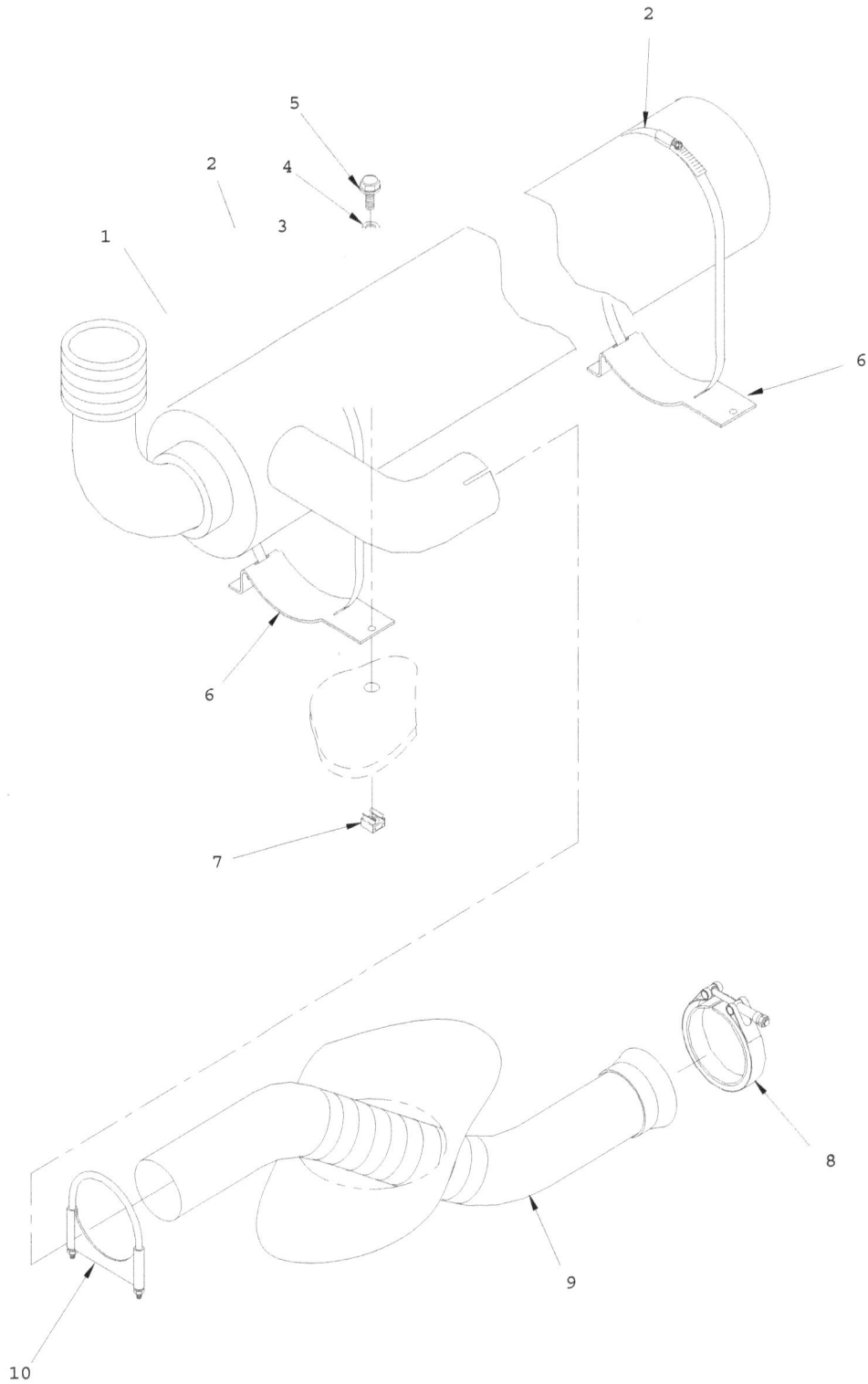

Figure 7. Air Intake and Exhaust System (Sheet 1 of 3)

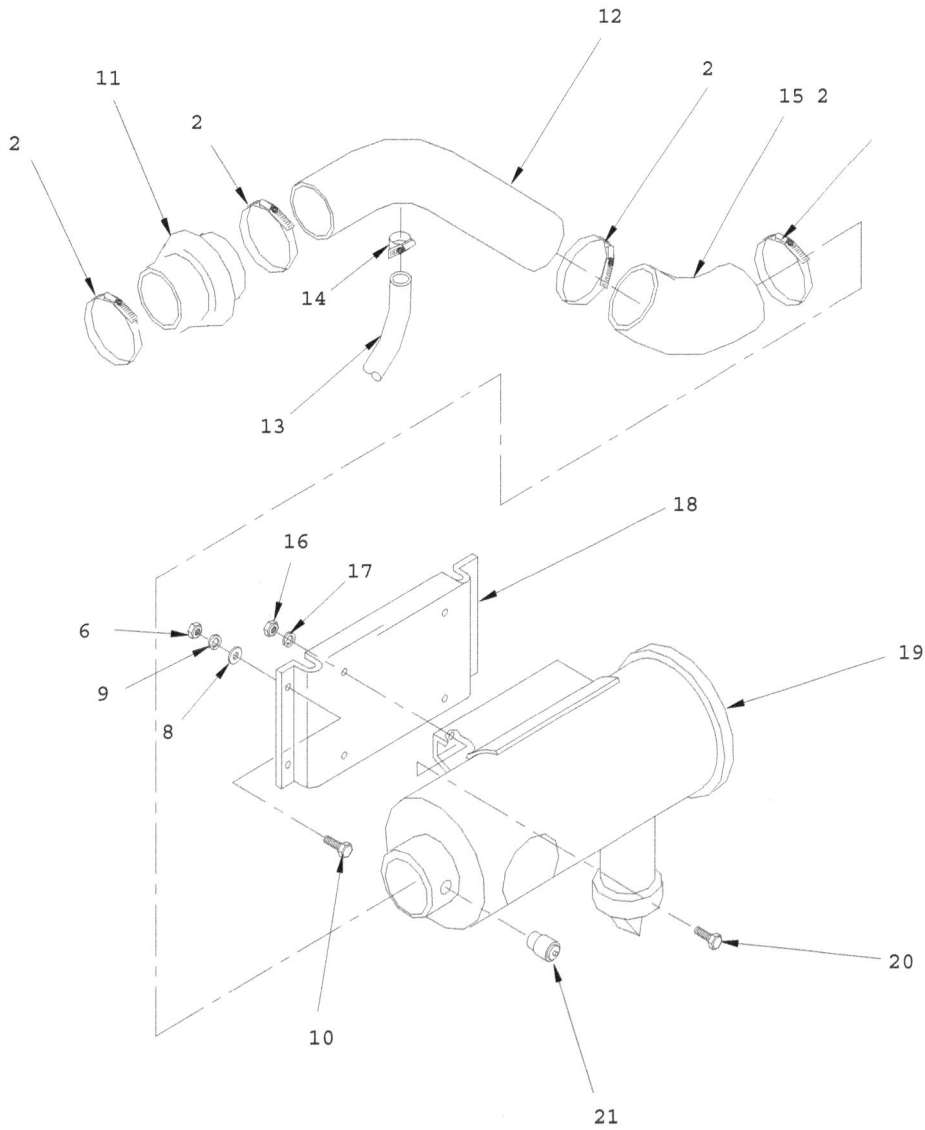

Figure 7. Air Intake and Exhaust System (Sheet 2 of 3)

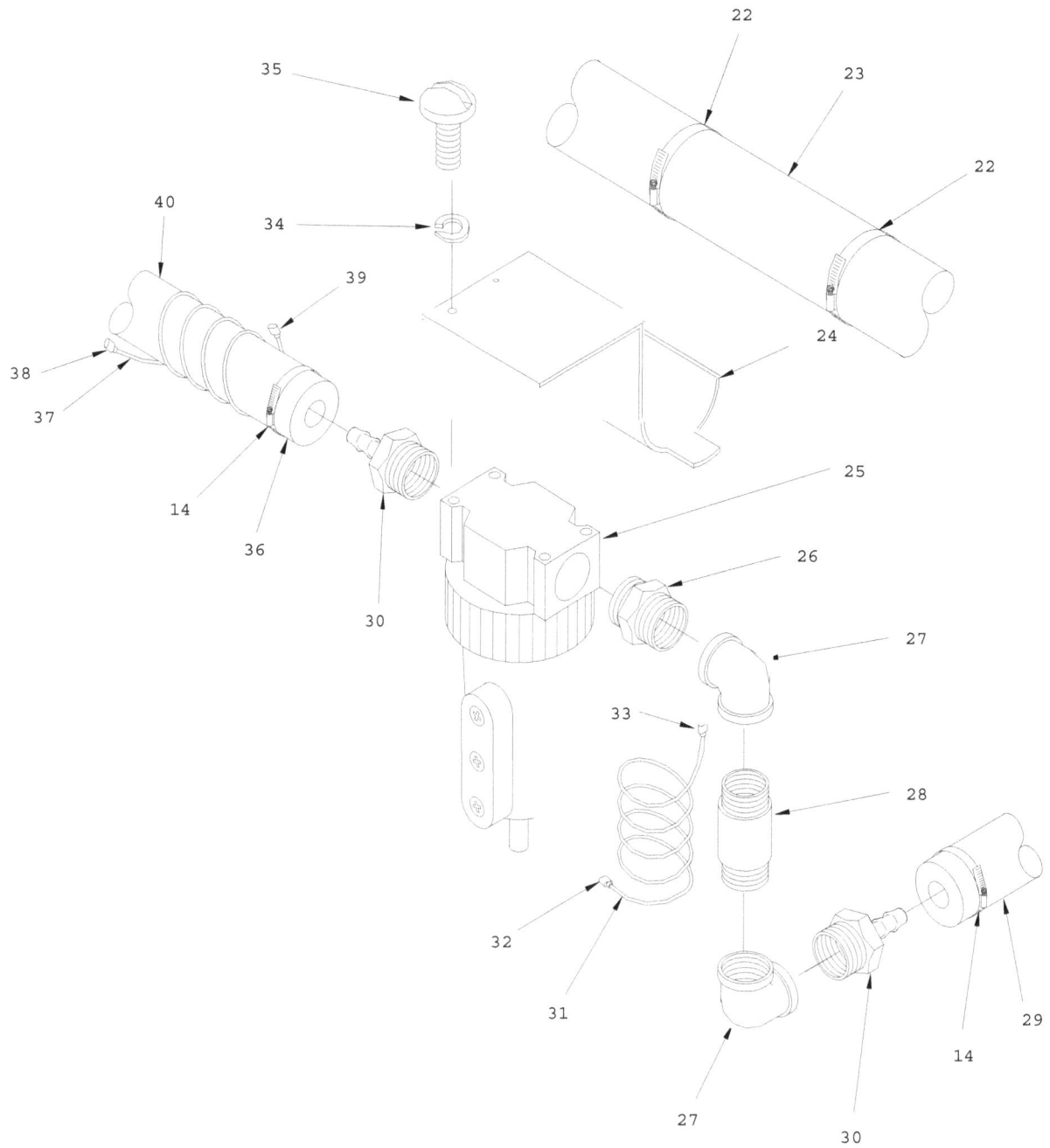

Figure 7. Air Intake and Exhaust System (Sheet 3 of 3)

(1) ITEM NO	(2) ARMY	AIR FORCE	USMC	(3) CAGEC	(4) SMR CODE PART NUMBER	(5) DESCRIPTION AND USABLE ON CODE (UOC)	(6) QTY
						GROUP 04 AIR INTAKE AND EXHAUST SYSTEM	
						FIGURE 7 AIR INTAKE AND EXHAUST SYSTEM	
1	PAOZZ	PAO00	PAOZZ	1DG36	400112	MUFFLER,EXHAUST........................1
2	PAOZZ	PAO00	PAOZZ	30554	88-20561-6	CLAMP,HOSE............................8
3	PAOZZ	PAO00	PAOZZ	96906	MS27183-10	WASHER,FLAT........................12
4	PAOZZ	PAO00	PAOZZ	97403	13230E6744-63	WASHER,FLAT........................12
5	PAOZZ	PAO00	PAOZZ	24617	274825	SCREW,CAP,HEXAGON H..................8
6	PAOZZ	PAO00	PAOZZ	18265	PPP20-6605	CLAMP,LOOP...........................1
7	PAOZZ	PAO00	PAOZZ	30554	88-21674-3	NUT,CAGE.............................8
8	PAOZZ	PAO00	PAOZZ	79260	35794	CLAMP,LOOP............................1
9	PAOZZ	PAO00	PAOZZ	1DG36	021034	PIPE,EXHAUST.........................1
10	PAOZZ	PAO00	PAOZZ	30554	96-23588	BRACKET,MOUNTING.....................1
11	PAOZZ	PAO00	PAOZZ	98441	38300-1-56-48RE	HOSE,PREFORMED.......................1
12	XBOZZ	XB	XBOZZ	30554	88-21968	TUBE ASSEMBLY,INTAK.................1
13	MOOZZ	MO00	MO000	30554	96-23681-2	HOSE,NONMETALLIC.....................1
14	PAOZZ	PAO00	PAOZZ	30554	88-20561-2	CLAMP,HOSE...........................3
15	PAOZZ	PAO00	PAOZZ	18265	P109331-016-700	HOSE,ELBOW...........................1
16	PAOZZ	PAO00	PAOZZ	30554	88-22790-3	NUT,PLAIN,HEXAGON....................4
17	PAOZZ	PAO00	PAOZZ	97403	13230E6744-65	WASHER,LOCK..........................4
18	XBOZZ	XB	XBOZZ	30554	88-21743	SUPPORT,AIR CLEANER..................1
19	XB000	XB	XB000	18265	FHG08-0511	CLEANER,AIR..........................1
						(SEE FIGURE 8 FOR PARTS BREAKDOWN)
20	PAOZZ	PAO00	PAOZZ	80204	B1821BH038C075N	SCREW,CAP,HEXAGON H..................4
21	PAOZZ	PAO00	PAOZZ	18265	RBX00-6379	INDICATOR,PRESSURE...................1
22	PAOZZ	PAO00	PAOZZ	30554	88-20561-3	CLAMP,HOSE...........................2
23	MOOZZ	MO00	MO000	30554	96-23705-3	INSULATION,PIPE......................V
						MAKE FROM P/N APT 11838 (OBSCO)
24	XBOZZ	XB	XBOZZ	30554	96-23672	BRACKET,MOUNTING.....................1
25	XB000	XB	XB000	16327	2Z329B	AIR CLEANER,INTAKE...................1
						(SEE FIGURE 9 FOR PARTS BREAKDOWN)
26	PAOZZ	PAO00	PAOZZ	50019	S1990	NIPPLE,PIPE..........................2
27	PAOZZ	PAO00	PAOZZ	3Z031	602350050	CONNECTOR,MULTIPLE...................3
28	PAOZZ	PAO00	PAOZZ	96906	MS51846-64	NIPPLE,TUBE..........................1
29	MOOZZ	MO00	MO000	30554	96-23681-1	HOSE,NONMETALLIC.....................1
						MAKE FROM P/N 160-12 (98441), 8.50 IN
30	PAOZZ	PAO00	PAOZZ	30554	88-21755-2	ADAPTER,STRAIGHT,PI..................2
31	PAOZZ	PAO00	PAOZZ	30554	96-23703	HEATER ASSEMBLY,ROP..................1
32	PAOZZ	PAO00	PAOZZ	56501	RA2573	TERMINAL,DISCONNECT..................1
33	PAOZZ	PAO00	PAOZZ	56501	RA25177	TERMINAL,QUICK DISC..................1
34	PAOZZ	PAO00	PAOZZ	97403	13230E6744-62	WASHER,LOCK-SPRING...................4
35	PAOZZ	PAO00	PAOZZ	45722	P-15121-64	SCREW,ASSEMBLED WAS..................4
36	MOOZZ	MO00	MO000	30554	96-23700-2	PIPE,FLEXIBLE........................V

(1)	(2)			(3)	(4)	(5)	(6)
					SMR CODE		
ITEM		AIR			PART	DESCRIPTION AND	
NO	ARMY	FORCE	USMC	CAGEC	NUMBER	USABLE ON CODE (UOC)	QTY

```
37    PAOZZ    PAOOO    PAOZZ    30554  96-23697      HEATER ASSEMBLY,ROP...................1
38 PAOZZ    PAOOO    PAOZZ 56501 RA2573 TERMINAL,DISCONNECT...................1
39 PAOZZ    PAOOO    PAOZZ 56501 RA25177 TERMINAL,QUICK DISC..................1 40    MOOZZ    MOOO
MOOOO    30554  96-23702-2      TAPE,FOIL............................V
                                              MAKE FROM P/N 1430 (52152)

                                     END OF FIGURE
```

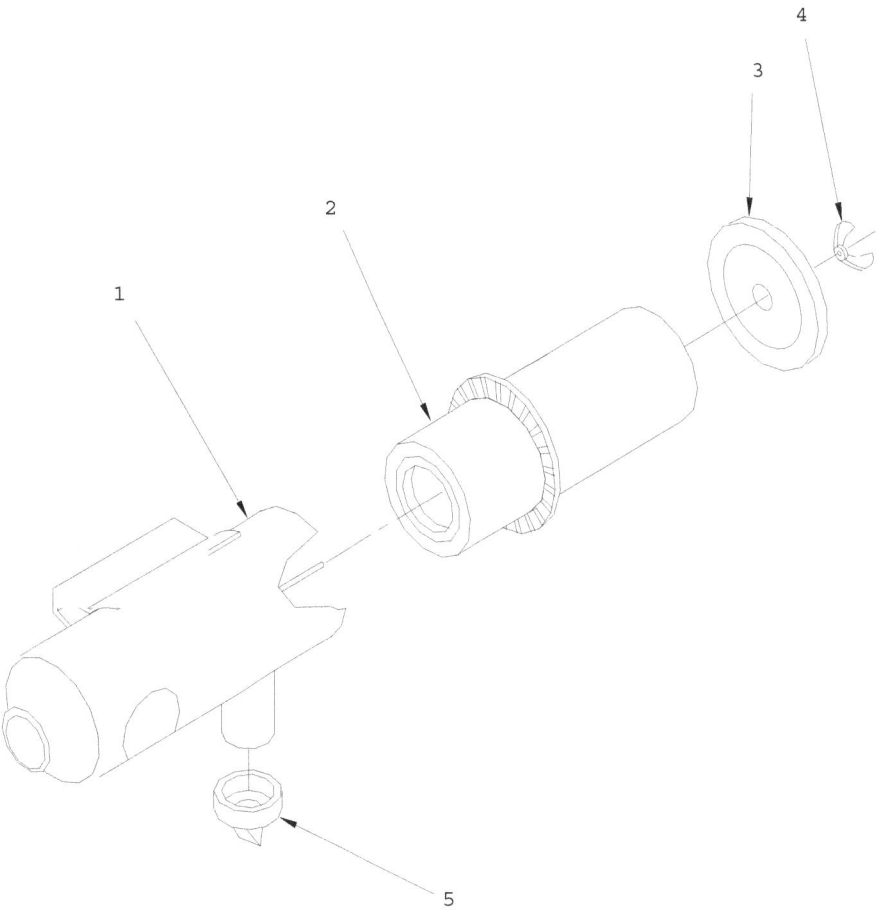

Figure 8. Air Cleaner Assembly

(1)		(2)		(3)	(4) SMR CODE	(5)	(6)
ITEM NO	ARMY	AIR FORCE	USMC	CAGEC	PART NUMBER	DESCRIPTION AND USABLE ON CODE (UOC)	QTY

GROUP 0402 AIR CLEANER ASSEMBLY
FIGURE 8 AIR CLEANER ASSEMBLY

1	XBOZZ	XB	XBOZZ	30554	88-21046	CANNISTER,AIR CLEAN...................1	
2	PAOZZ	PAOOO	PAOZZ	18265	P18-2059	FILTER ELEMENT,INTA...................1	
3	XBOZZ	XB	XBOZZ	18265	P11-9711	COVER,FLUID FILTER....................1	
4	XBOZZ	XB	XBOZZ	18265	P10-1870	NUT,PLAIN,WING........................1	
5	PAOZZ	PAOOO	PAOZZ	18265	P10-6593	VALVE,VACUATOR........................1	

END OF FIGURE

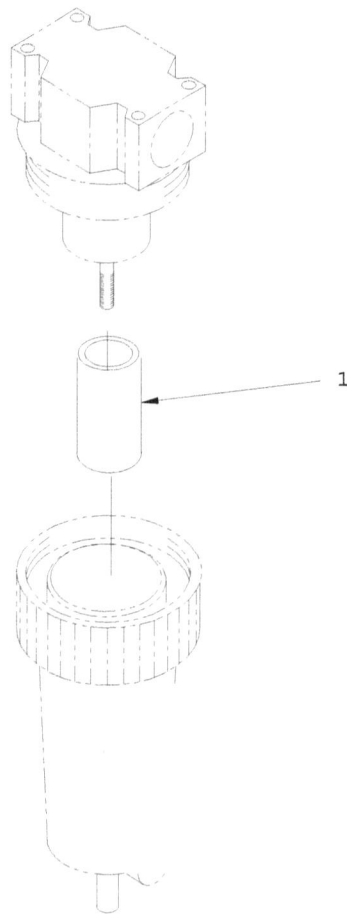

Figure 9. Crankcase Breather Filter Assembly

(1)		(2)		(3)	(4)	(5)	(6)
					SMR CODE		
ITEM		AIR			PART	DESCRIPTION AND	
NO	ARMY	FORCE	USMC	CAGEC	NUMBER	USABLE ON CODE (UOC)	QTY

GROUP 0403 CRANKCASE BREATHER FILTER ASSEMBLY
FIGURE 9 CRANKCASE BREATHER FILTER ASSEMBLY

| 1 | PAOZZ | PAO00 | PAOZZ | 16327 | 1R424A | PARTS KIT,AIR FILTER,WITH O-RING......1 | |

END OF FIGURE

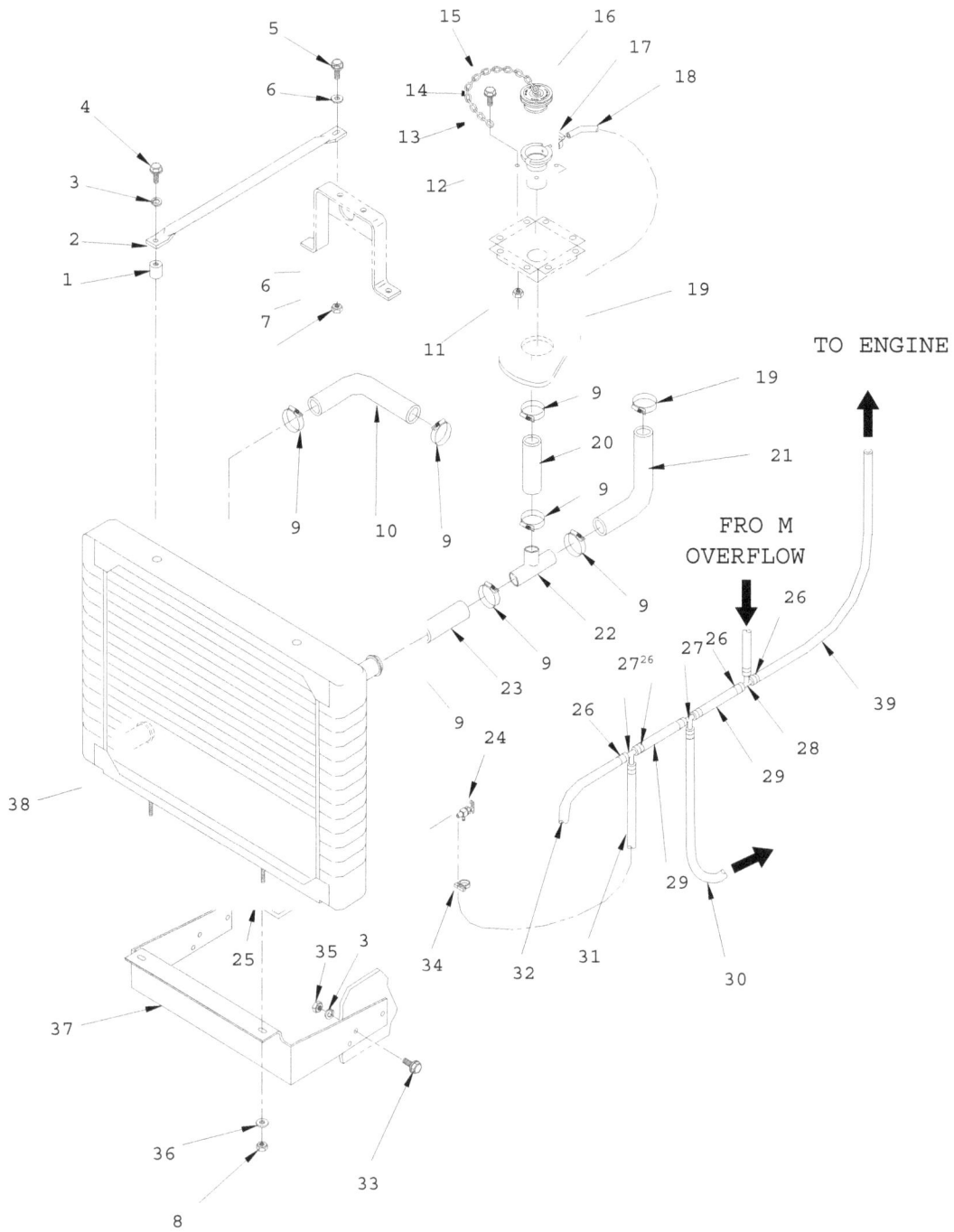

Figure 10. Engine Cooling System (Sheet 1 of 3)

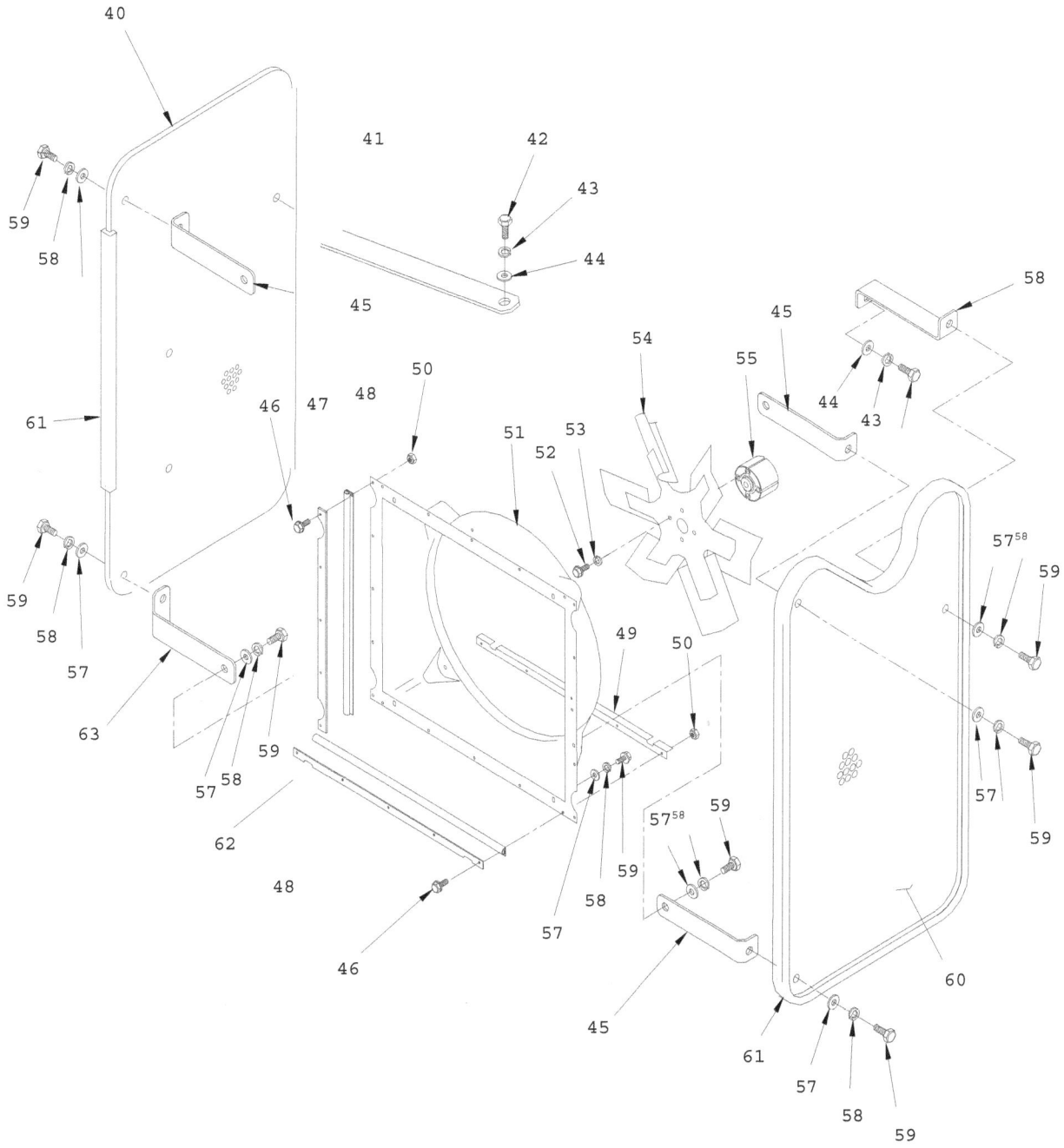

Figure 10. Engine Cooling System (Sheet 2 of 3)

65

34

64

FRO M
SHEET

HOT

COLD

34

18
REF

59

57

58

66

67

P/O67

42

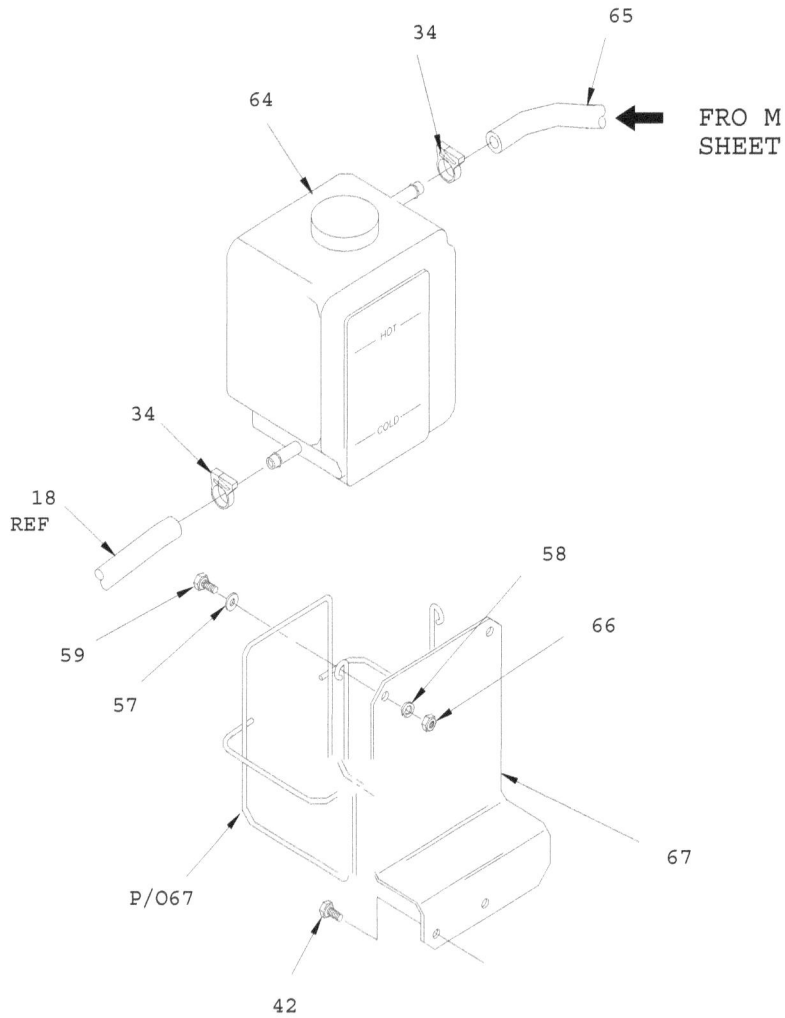

Figure 10. Engine Cooling System (Sheet 3 of 3)

(1)		(2)		(3)	(4) SMR CODE PART	(5)	(6)
ITEM NO	ARMY	AIR FORCE	USMC	CAGEC	NUMBER	DESCRIPTION AND USABLE ON CODE (UOC)	QTY

GROUP 05 ENGINE COOLING SYSTEM
FIGURE 10 ENGINE COOLING SYSTEM

(1)		(2)		(3)	(4)	(5)	(6)
1	XB0ZZ	XB	XB0ZZ	30554	88-22145	SPACER,TIE ROD.........................1	
2	XB0ZZ	XB	XB0ZZ	30554	88-21711	TIE ROD,SUPPORT,RAD..................1	
3	PA0ZZ	PA000	PA0ZZ	97403	13230E6744-65	WASHER,LOCK.........................1	
4	PA0ZZ	PA000	PA0ZZ	80204	B1821BH038C150N	SCREW,CAP,HEXAGON H..................1	
5	PA0ZZ	PA000	PA0ZZ	80204	B1821BH031C100N	BOLT,MACHINE.........................2	
6	PA0ZZ	PA000	PA0ZZ	96906	MS51412-25	WASHER,FLAT........................4	
7	PA0ZZ	PA000	PA0ZZ	97403	13230E6744-64	WASHER,LOCK-SPRING..................2	
8	PA0ZZ	PA000	PA0ZZ	30554	88-22790-2	NUT,HEXAGON.........................2	
9	PA0ZZ	PA000	PA0ZZ	30554	88-20561-4	CLAMP,HOSE.........................8	
10	PA0ZZ	PA000	PA0ZZ	30554	96-23551	HOSE,NONMETALLIC....................1	
11	PA0ZZ	PA000	PA0ZZ	19617	HW310	NUT,PLAIN,ASSEMBLED.................4	
12	XB0ZZ	XB	XB0ZZ	41197	3A43788B	NECK,FILLER,RADIATO.................1	
13	PA0ZZ	PA000	PA0ZZ	32529	53	HOOK,CHAIN,S.........................2	
14	PA0ZZ	PA000	PA0ZZ	19207	12325869	BOLT,MACHINE.........................4	
15	PA0ZZ	PA000	PA0ZZ	81343	ASTM A 466	CHAIN,ZINC PLATED...................1	
16	PA0ZZ	PA000	PA0ZZ	30554	96-23710	CAP,FILLER OPENING..................1	
17	PA0ZZ	PA000	PA0ZZ	30554	88-20561-1	CLAMP,HOSE.........................1	
18	M00ZZ	M000	M0000	30554	96-23610-1	HOSE,NONMETALLIC....................1 MAKE FROM P/N 55-1986-19 (99739)	
19	PA0ZZ	PA000	PA0ZZ	70485	804	GROMMET,NONMETALLIC.................1	
20	M00ZZ	M000	M0000	30554	96-23599	HOSE,RUBBER.........................1 MAKE FROM SAE J20R5, CLASS D-2 (81348)	
21	PA0ZZ	PA000	PA0ZZ	30554	96-23550	HOSE,NONMETALLIC....................1	
22	PA0ZZ	PA000	PA0ZZ	30554	96-23652	TEE,HOSE............................1	
23	PA0ZZ	PA000	PA0ZZ	30554	96-23595	HOSE,NONMETALLIC....................1	
24	PA0ZZ	PA000	PA0ZZ	93061	NV108P-4	VALVE,DRAIN.........................1	
25	XB0ZZ	XB	XB0ZZ	30554	88-22703	SPACER,SPECIAL SHAP.................1	
26	PA0ZZ	PA000	PA0ZZ	28520	2314	CLAMP,HOSE.........................6	
27	PA0ZZ	PA000	PA0ZZ	79470	1908	TEE,HOSE............................2	
28	PA0ZZ	PA000	PA0ZZ	79470	1945	TEE,HOSE............................1	
29	M00ZZ	M000	M0000	30554	96-23610-2	HOSE,NONMETALLIC....................2 MAKE FROM P/N 55-1986-19 (99739)	
30	M00ZZ	M000	M0000	30554	96-23609-5	HOSE,NONMETALLIC....................1 MAKE FROM P/N 483666 (03380)	
31	M00ZZ	M000	M0000	30554	96-23609-2	HOSE,NONMETALLIC....................1 MAKE FROM P/N 483666 (03380)	
32	M00ZZ	M000	M0000	30554	96-23610-7	HOSE,NONMETALLIC....................1 MAKE FROM P/N 55-1986-19 (99739)	
33	PA0ZZ	PA000	PA0ZZ	80204	B1821BH038C125N	SCREW,CAP,HEXAGON H..................6	
34	PA0ZZ	PA000	PA0ZZ	28520	2326	CLAMP,HOSE.........................3	
35	PA0ZZ	PA000	PA0ZZ	30554	88-22790-3	NUT,PLAIN,HEXAGON...................6	
36	PA0ZZ	PA000	PA0ZZ	96906	MS51412-2	WASHER,FLAT........................2	

(1)	(2)			(3)	(4)	(5)	(6)
		AIR			SMR CODE PART	DESCRIPTION AND	
ITEM NO	ARMY	FORCE	USMC	CAGEC	NUMBER	USABLE ON CODE (UOC)	QTY

37	XB0ZZ	XB	XB0ZZ	30554	88-22049	SUPPORT,RADIATOR......................1	
38	PA0ZZ	PA000	PA0ZZ	30554	96-23547	RADIATOR,ENGINE COO...................1	
39	M00ZZ	M000	M0000	30554	96-23690-7	HOSE,NONMETALLIC......................1	
						MAKE FROM P/N 208-4 (98441)	
40	PA0ZZ	PA000	PA0ZZ	30554	96-23558	GUARD,FAN IMPELLER....................1	
41	XB0ZZ	XB	XB0ZZ	30554	96-23605	BRACKET,MOUNTING......................1	
42	PA0ZZ	PA000	PA0ZZ	48370	933-M10	SCREW,CAP,HEXAGON H...................4	
43	PA0ZZ	PA000	PA0ZZ	30554	96-23613-2	WASHER,SPLIT..........................4	
44	PA0ZZ	PA000	PA0ZZ	15526	125AZY-10M	WASHER,FLAT...........................4	
45	XB0ZZ	XB	XB0ZZ	30554	96-23687	BRACKET,MOUNTING......................3	
46	PA0ZZ	PA000	PA0ZZ	30554	88-20565-1	SCREW,MACHINE.........................20	
47	XB0ZZ	XB	XB0ZZ	30554	88-22717	SUPPORT,SIDE,RADIAT...................2	48
	PA0ZZ	PA000	PA0ZZ	30554	88-22712	SEAL,PLAIN,CUT TO LENGTH..............V	
49	XB0ZZ	XB	XB0ZZ	30554	88-22720	STIFFENER,SHROUD......................2	
50	PA0ZZ	PA000	PA0ZZ	019L2	21NTE82	NUT,SELF-LOCKING......................20	
51	PB0ZZ	PB000	PB0ZZ	30554	88-21879	SHROUD,FAN,RADIATOR...................1	
52	PA0ZZ	PA000	PA0ZZ	15526	933-ZY-M8-65	BOLT,MACHINE..........................4	
53	PA0ZZ	PA000	PA0ZZ	15526	127BZY-8M	WASHER,SPLIT..........................4	
54	PA0ZZ	PA000	PA0ZZ	45454	320777	IMPELLER,FAN,AXIAL....................1	
55	XB0ZZ	XB	XB0ZZ	31874	7812-01	SPACER,FAN............................1	
56	XB0ZZ	XB	XB0ZZ	30554	96-23559	SUPPORT,STRUCTURAL....................1	
57	PA0ZZ	PA000	PA0ZZ	96906	MS27183-10	WASHER,FLAT...........................15	
58	PA0ZZ	PA000	PA0ZZ	97403	13230E6744-63	WASHER,FLAT...........................15	
59	PA0ZZ	PA000	PA0ZZ	80204	B1821BH025C075N	SCREW,CAP,HEXAGON H...................15	
60	PA0ZZ	PA000	PA0ZZ	30554	96-23582-1	GUARD,FAN IMPELLER....................1	61
	PA000	PA0ZZ		1E045	TL62B7X1/16	MOLDING,PLASTIC.......................V	
62	XB0ZZ	XB	XB0ZZ	30554	88-22719	SUPPORT,SEAL,RADIAT...................2	
63	XB0ZZ	XB	XB0ZZ	30554	96-23602	BRACKET,MOUNTING......................1	
64	PA0ZZ	PA000	PA0ZZ	30554	88-20468	TANK,RADIATOR,OVERF...................1	65 M00ZZ M000
	M0000			30554	96-23610-5	HOSE,NONMETALLIC......................V	
						MAKE FROM P/N 55-1986-19 (99739)	
66	PA0ZZ	PA000	PA0ZZ	30554	88-22790-1	NUT...................................2	
67	XB0ZZ	XB	XB0ZZ	30554	96-23529	BRACKET,MOUNTING......................1	

END OF FIGURE

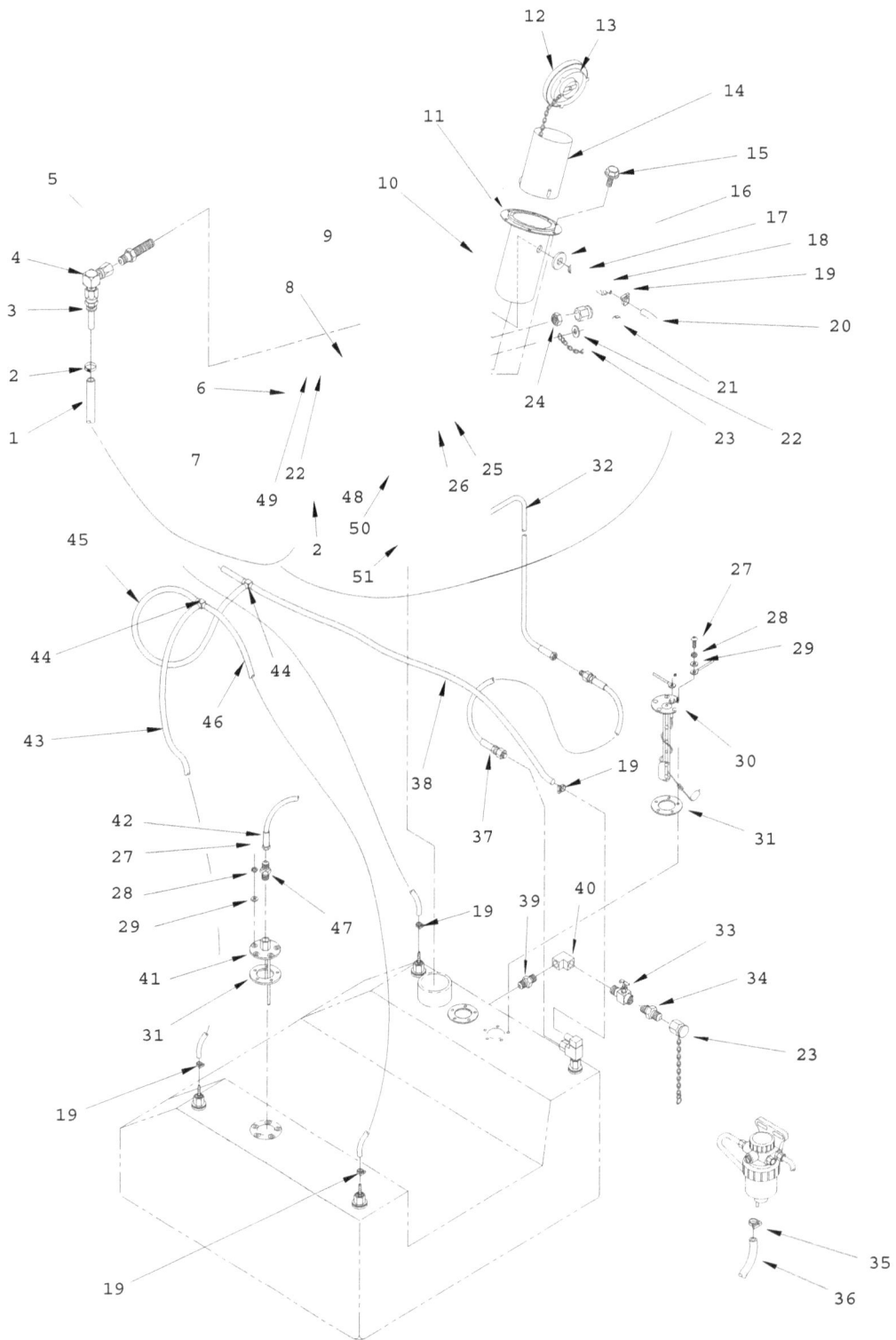

Figure 11. Fuel System (Sheet 1 of 3)

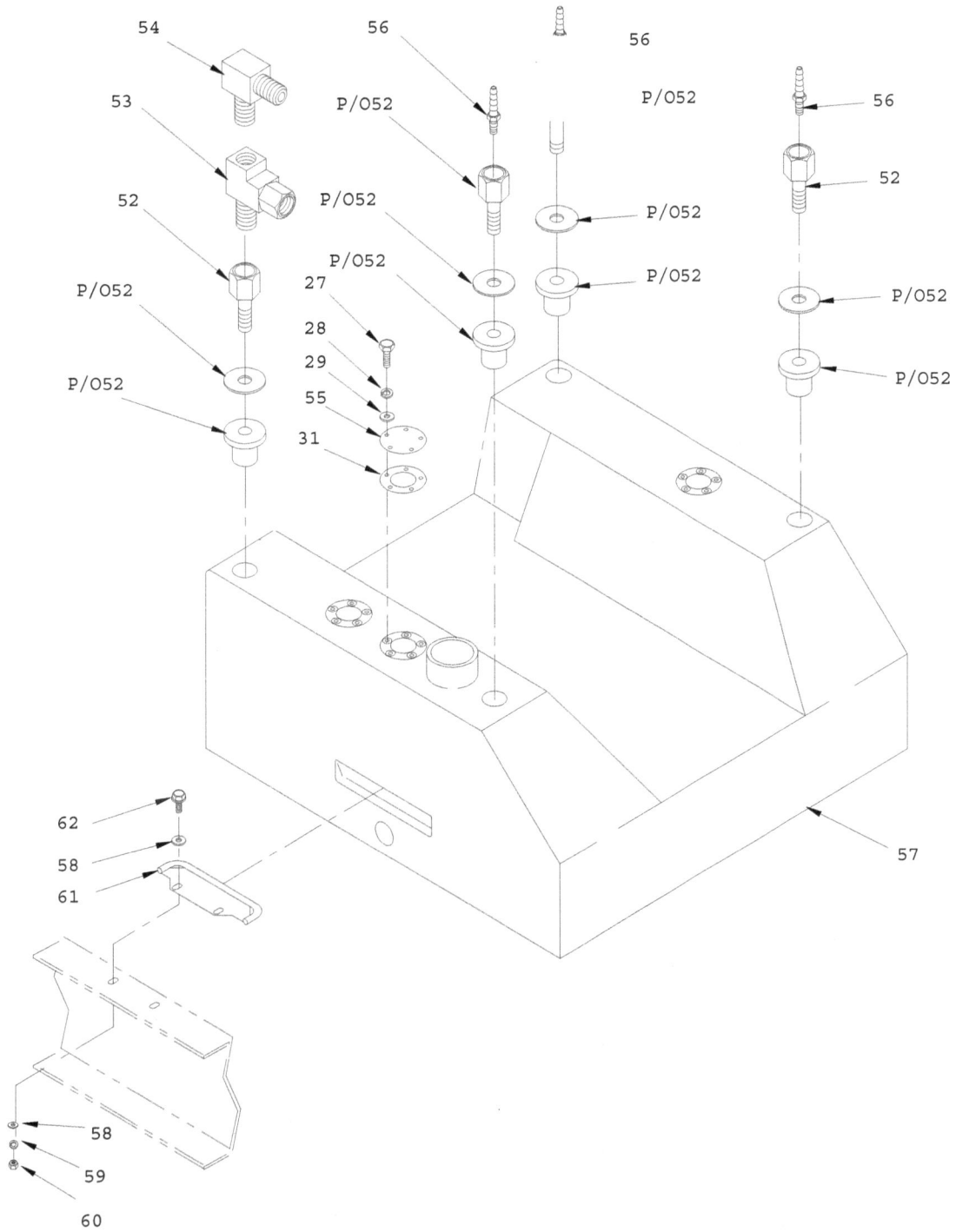

Figure 11. Fuel System (Sheet 2 of 3)

VIEW
ROTATED
180°

64

21

63

49

21

65

P/O65

P/O65

P/O65

69

66

63
22
49

P/O65

68

67

TO WIRING

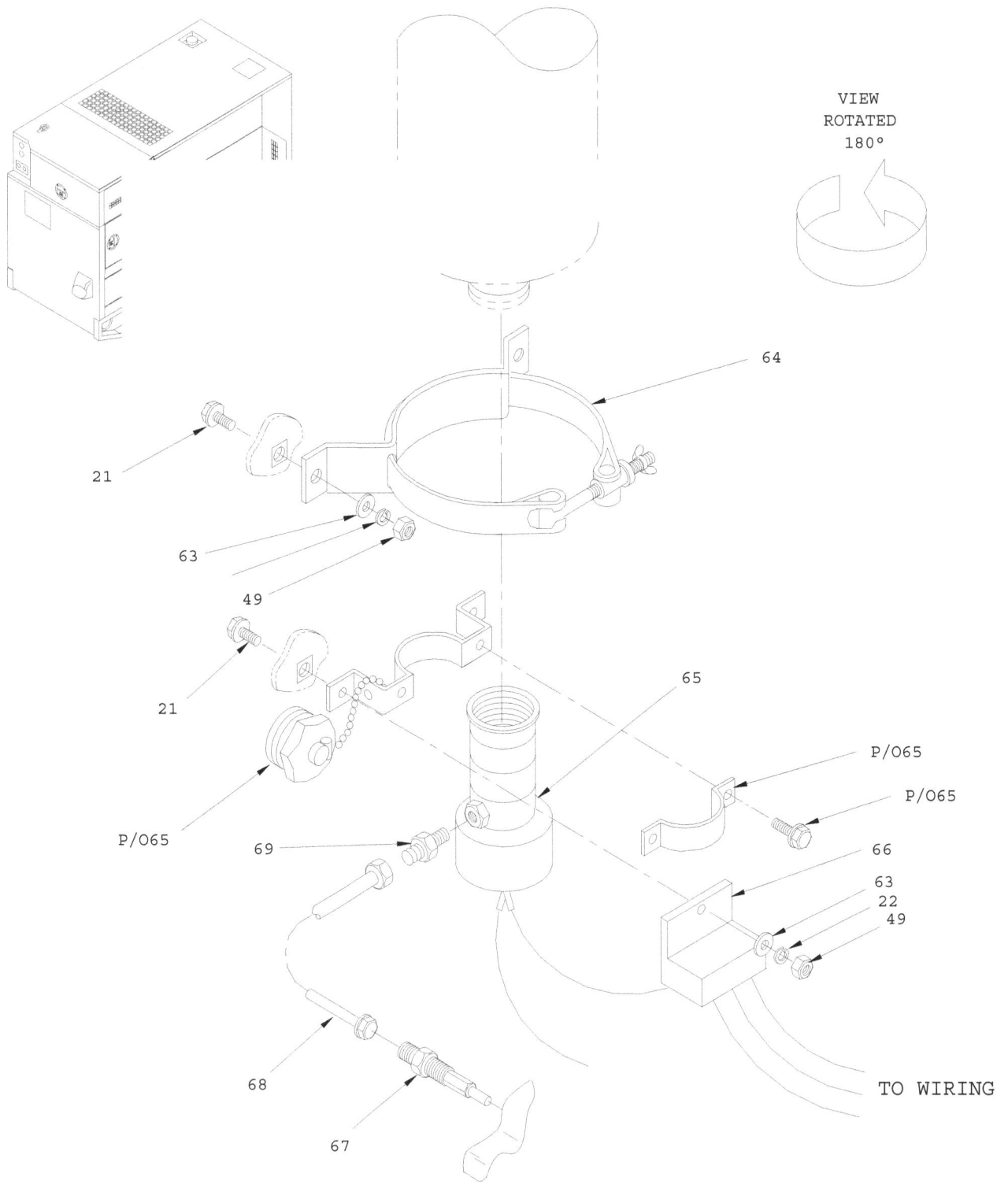

Figure 11. Fuel System (Sheet 3 of 3)

(1) ITEM NO	(2) ARMY	AIR FORCE	USMC	(3) CAGEC	SMR CODE PART NUMBER	(5) DESCRIPTION AND USABLE ON CODE (UOC)	(6) QTY
						GROUP 06 FUEL SYSTEM FIGURE 11 FUEL SYSTEM	
1	M00ZZ	M000	M0000	98441	96-23690-4	HOSE,NONMETALLIC.....................1 MAKE FROM P/N 208-5 (98441), 11.0 IN	
2	PA0ZZ	PA000	PA0ZZ	30554	88-20561-1	CLAMP,HOSE............................2	
3	PA0ZZ	PA000	PA0ZZ	5U990	205-605	ADAPTER,STRAIGHT,TU...................1	
4	PA0ZZ	PA000	PA0ZZ	96906	MS51521B5	ELBOW,TUBE............................1	
5	PA0ZZ	PA000	PA0ZZ	81343	5-5-070601	NIPPLE,TUBE...........................1	
6	XB0ZZ	XB	XB0ZZ	30554	88-22111	TUBE ASSEMBLY,OUTLE...................1	
7	PA0ZZ	PA000	PA0ZZ	79470	00904F-504-J04-0	HOSE ASSEMBLY,NONME...................1	
8	PA000	PA000	PA000	30554	88-22546	PUMP,FUEL,CAM ACTIV...................1 (SEE FIGURE 12 FOR PARTS BREAKDOWN)	
9	XB0ZZ	XB	XB0ZZ	30554	96-23577	PANEL,FILLER NECK.....................1	
10	XB0ZZ	XB	XB0ZZ	79470	C3059X2	SEAL NUT,PIPE.........................1	
11	XB0ZZ	XB	XB0ZZ	30654	88-21892	FILLER NECK,LIQUID....................1	
12	PA0ZZ	PA000	PA0ZZ	96906	MS35645-1	CAP,FILLER OPENING....................1	
13	PA0ZZ	PA000	PA0ZZ	78225	GAR-0661-1	GASKET................................1	
14	PA0ZZ	PA000	PA0ZZ	30554	88-21893	FILLER,TUBE,FUEL TA...................1	
15	PA0ZZ	PA000	PA0ZZ	30554	88-20260-23	SCREW,CAP,HEXAGON H...................6	
16	PA0ZZ	PA000	PA0ZZ	96906	MS51412-8	WASHER,FLAT...........................1	
17	XB0ZZ	XB	XB0ZZ	81343	4-2-140137	NIPPLE,HEX,REDUCER....................1	
18	PA0ZZ	PA000	PA0ZZ	36378	1001547-01	ADAPTER,STRAIGHT,PI...................1	
19	PA0ZZ	PA000	PA0ZZ	19207	11608950-18	CLAMP,HOSE...........................11	
20	M00ZZ	M000	M0000	30554	96-23690-1	HOSE,NONMETALLIC......................1 MAKE FROM P/N 208-4 (98441)	
21	PA0ZZ	PA000	PA0ZZ	30554	88-20260-31	BOLT,MACHINE..........................5	
22	PA0ZZ	PA000	PA0ZZ	97403	13230E6744-63	WASHER,FLAT...........................8	
23	PA0ZZ	PA000	PA0ZZ	93742	69-539-2	CAP,TUBE..............................2	
24	PA0ZZ	PA000	PA0ZZ	81343	5-5-070118	NUT,SELF-LOCKING,CL...................1	
25	PA0ZZ	PA000	PA0ZZ	78189	511-081800-00	NUT...................................6	
26	PA0ZZ	PA000	PA0ZZ	30554	88-20561-7	CLAMP,HOSE............................1	
27	PA0ZZ	PA000	PA0ZZ	30554	88-22793-9	SCREW,MACHINE........................15	
28	PA0ZZ	PA000	PA0ZZ	97403	13230E6744-62	WASHER,LOCK-SPRING...................15	
29	PA0ZZ	PA000	PA0ZZ	96906	MS51412-2	WASHER,FLAT..........................15	
30	PA0ZZ	PA000	PA0ZZ	06555	127678	TRANSMITTER,LIQUID....................1	
31	PA0ZZ	PA000	PA0ZZ	30554	88-20286	GASKET................................4	
32	PA0ZZ	PA000	PA0ZZ	30554	96-23581-2	TUBE ASSEMBLY,ME......................1	
33	PA0ZZ	PA000	PA0ZZ	93061	NV108P-4	COCK,SHUTOFF,SCREW....................1	
34	PA0ZZ	PA000	PA0ZZ	81343	4-5-070102	ADAPTER,PRESSURE FU...................1	
35	PA0ZZ	PA000	PA0ZZ	28520	2326	CLAMP,HOSE............................1	
36	M00ZZ	M000	M0000	30554	96-23609-5	HOSE,NONMETALLIC......................1 MAKE FROM P/N 483666 (03380)	
37	PA0ZZ	PA000	PA0ZZ	98441	A3411-54	HOSE ASSEMBLY,NONME...................1	

(1)	(2)			(3)	(4)	(5)	(6)
					SMR CODE		
ITEM	AIR				PART	DESCRIPTION AND	
NO	ARMY	FORCE	USMC	CAGEC	NUMBER	USABLE ON CODE (UOC)	QTY
38	M0OZZ	MO00	M0O00	30554	96-23690-5	HOSE,NONMETALLIC.....................1	
						MAKE FROM P/N 208-4 (98441)	
39	PAOZZ	PA000	PAOZZ	30327	112B1-4IN	NIPPLE,PIPE..........................1	
40	PAOZZ	PA000	PAOZZ	21450	444069	ELBOW,PIPE...........................1	
41	PAOZZ	PA000	PAOZZ	30554	88-21851	TUBE ASSEMBLY,METAL..................1	
42	PAOZZ	PA000	PAOZZ	30554	96-23593-2	HOSE ASSEMBLY,NONME..................1	
43	M0OZZ	MO00	M0O00	30554	96-23690-3	HOSE,NONMETALLIC.....................1	
						MAKE FROM P/N 208-4 (98441)	
44	PAOZZ	PA000	PAOZZ	30554	88-21883	TEE,HOSE.............................2	
45	M0OZZ	MO00	M0O00	30554	96-23690-10	HOSE,NONMETALLIC.....................1	
						MAKE FROM P/N 208-4 (98441)	
46	M0OZZ	MO00	M0O00	30554	96-23690-7	HOSE,NONMETALLIC.....................1	
						MAKE FROM P/N 208-4 (98441)	
47	PAOZZ	PA000	PAOZZ	81343	4-4-070102	ELBOW,PIPE TO HOSE...................1	
48	PAOZZ	PA000	PAOZZ	72850	479735	FILTER BODY,FLUID....................1	
49	PAOZZ	PA000	PAOZZ	30554	88-22790-1	NUT..................................6	
50	PAOZZ	PA000	PAOZZ	30554	88-22068	HOSE,PREFORMED.......................1	
51	PAOZZ	PA000	PAOZZ	30554	88-20561-6	CLAMP,HOSE...........................1	
52	PAOZZ	PA000	PAOZZ	30554	88-20049	ADAPTER,PRESSURE FU..................5	
53	PAFZZ	PA000	PAFZZ	81343	4-4-4-140424	TEE,PIPE.............................1	
54	PAFZZ	PA000	PAFZZ	79470	1069X6	ELBOW,PIPE TO HOSE...................1	
55	XBFZZ	XB	XBFZZ	30554	96-23578	PLATE,METAL..........................1	
56	PAFZZ	PA000	PAFZZ	93061	125HBL-4-4	ADAPTER,STRAIGHT,PI..................3	
57	PAFZZ	PA000	PAFZZ	30554	88-21044	TANK,FUEL,ENGINE.....................1	
58	PAFZZ	PA000	PAFZZ	30554	88-20564-12	WASHER,FLAT..........................8	
59	PAFZZ	PA000	PAFZZ	97403	13230E6744-64	WASHER,LOCK-SPRING...................4	
60	PAFZZ	PA000	PAFZZ	30554	88-22790-2	NUT,HEXAGON..........................4	
61	XBFZZ	XB	XBFZZ	30554	88-21854	HOLDOWN ASSEMBLY,FU..................2	
62	PAFZZ	PA000	PAFZZ	80204	B1821BH031C100N	BOLT,MACHINE.........................4	
63	PAOZZ	PA000	PAOZZ	96906	MS27183-10	WASHER,FLAT..........................4	
64	XB0ZZ	XB	XB0ZZ	61112	QS-1189	CLAMP,LOOP...........................1	
65	PA000	PA000	PA000	30554	88-22553	SOLENOID ASSEMBLY....................1	
						(SEE FIGURE 13 FOR PARTS BREAKDOWN)	
66	PAOZZ	PA000	PAOZZ	61529	CB1AF-M-24V	RELAY,ELECTROMAGNET..................1	
67	PAOZZ	PA000	PAOZZ	50999	JNZ20S0S	INJECTOR ASSEMBLY,F..................1	
68	PAOZZ	PA000	PAOZZ	83616	TD2-03420	TUBING,NONMETALLIC...................1	
69	PAOZZ	PA000	PAOZZ	23040	C391674	ADAPTER,STRAIGHT,PI..................1	

END OF FIGURE

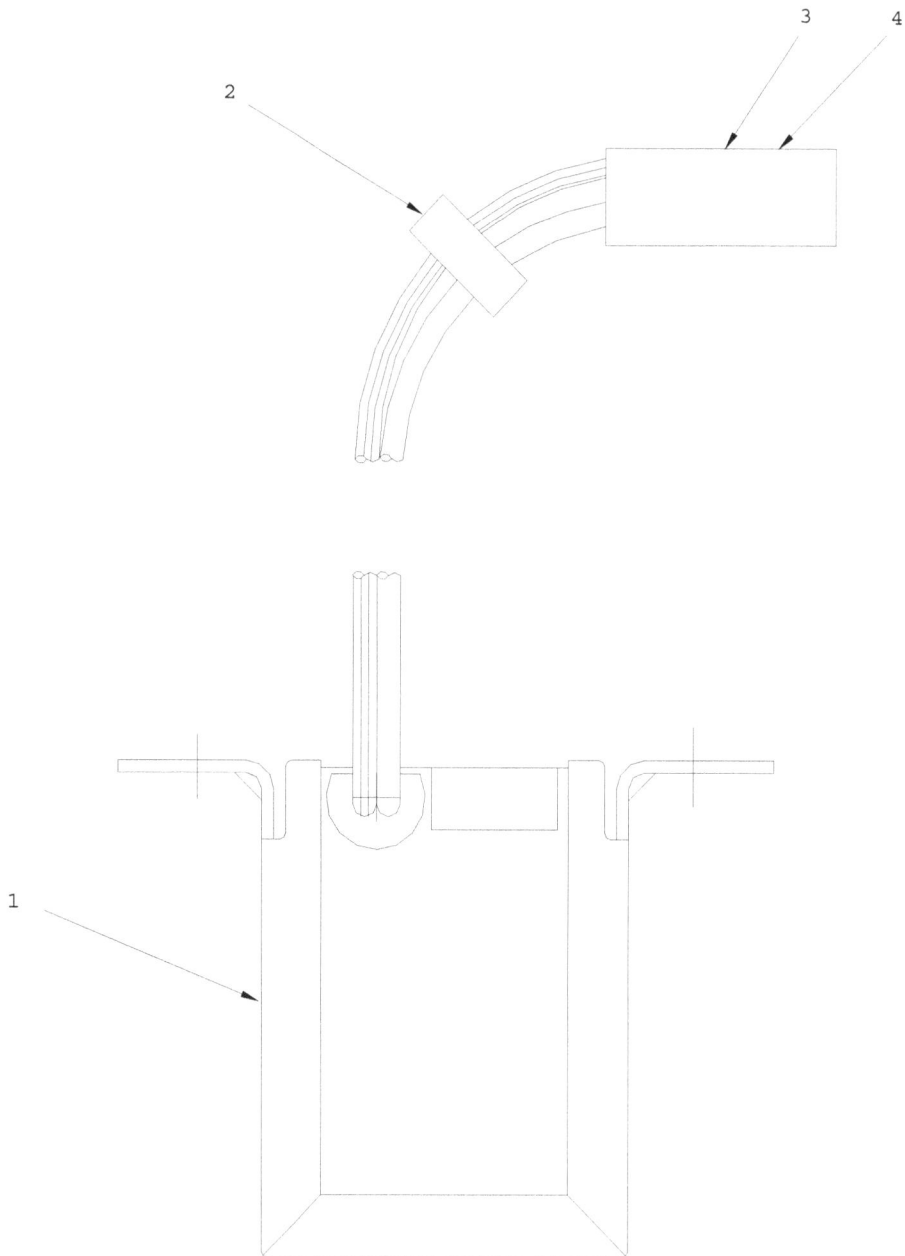

Figure 12. Auxiliary Fuel Pump

(1)		(2)		(3)	(4)	(5)	(6)
					SMR CODE		
ITEM		AIR			PART	DESCRIPTION AND	
NO	ARMY	FORCE	USMC	CAGEC	NUMBER	USABLE ON CODE (UOC)	QTY

GROUP 0602 AUXILIARY FUEL PUMP
FIGURE 12 AUXILIARY FUEL PUMP

1	PAOZZ	PAO00	PAOZZ	72850	40194	PUMP,FUEL,ELECTRICA 1
2	PAOZZ	PAO00	PAOZZ	06383	PLT2S	STRAP,TIEDOWN,ELECT....................1
3	PAOZZ	PAO00	PAOZZ	63123	35-000252-001	CONTACT,ELECTRICAL....................2
4	PAOZZ	PAO00	PAOZZ	27264	03-09-2022	CONNECTOR,PLUG,ELEC...................1

END OF FIGURE

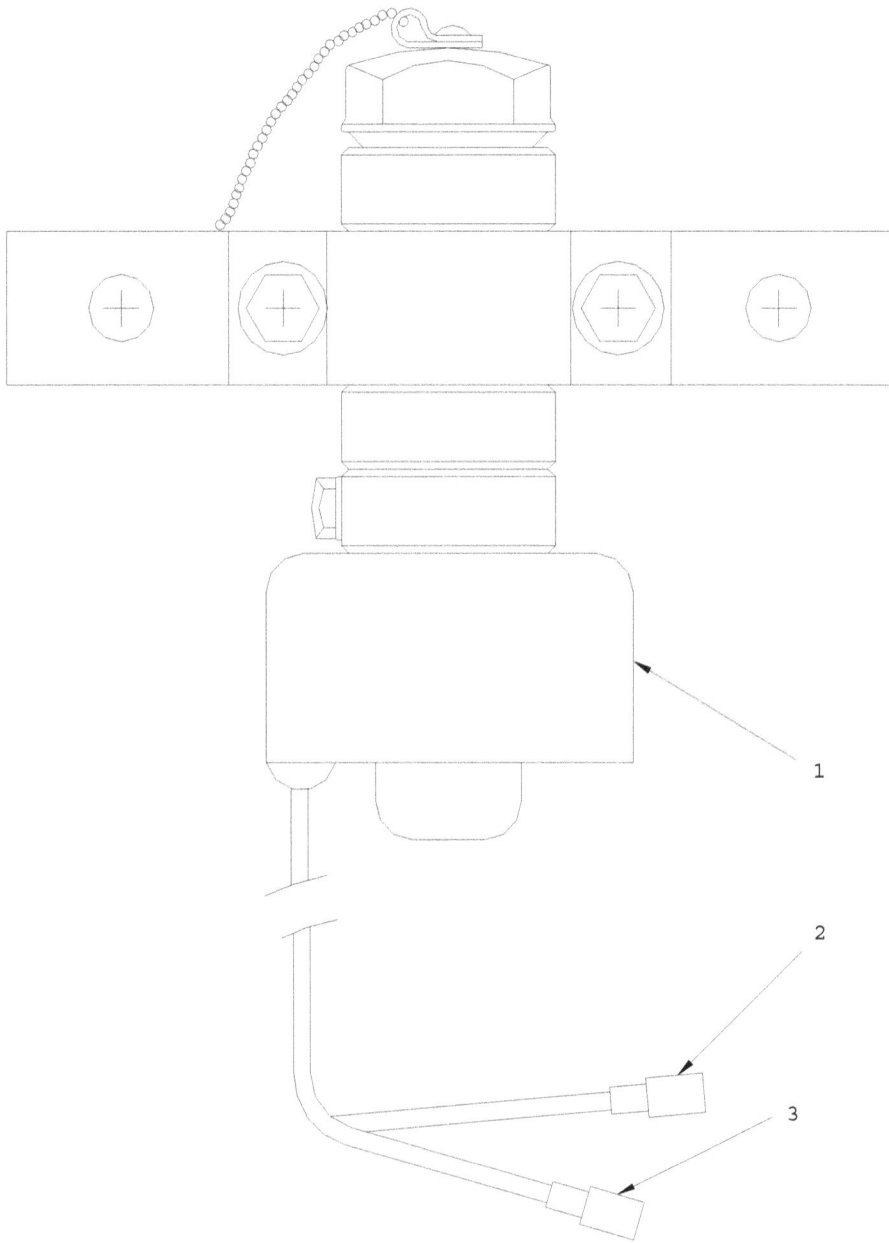

Figure 13. Ether Solenoid Valve

(1)		(2)		(3)	(4)	(5)	(6)
					SMR CODE		
ITEM		AIR			PART	DESCRIPTION AND	
NO	ARMY	FORCE	USMC	CAGEC	NUMBER	USABLE ON CODE (UOC)	QTY

GROUP 0608 ETHER SOLENOID VALVE
FIGURE 13 ETHER SOLENOID VALVE

1	PAOZZ	PAOOO	PAOZZ	78280	QS-4-2TC	VALVE,SOLENOID........................1
2	PAOZZ	PAOOO	PAOZZ	56501	RA2573	TERMINAL,DISCONNECT...................1
3	PAOZZ	PAOOO	PAOZZ	56501	RA2573	TERMINAL,QUICK DISC...................1

END OF FIGURE

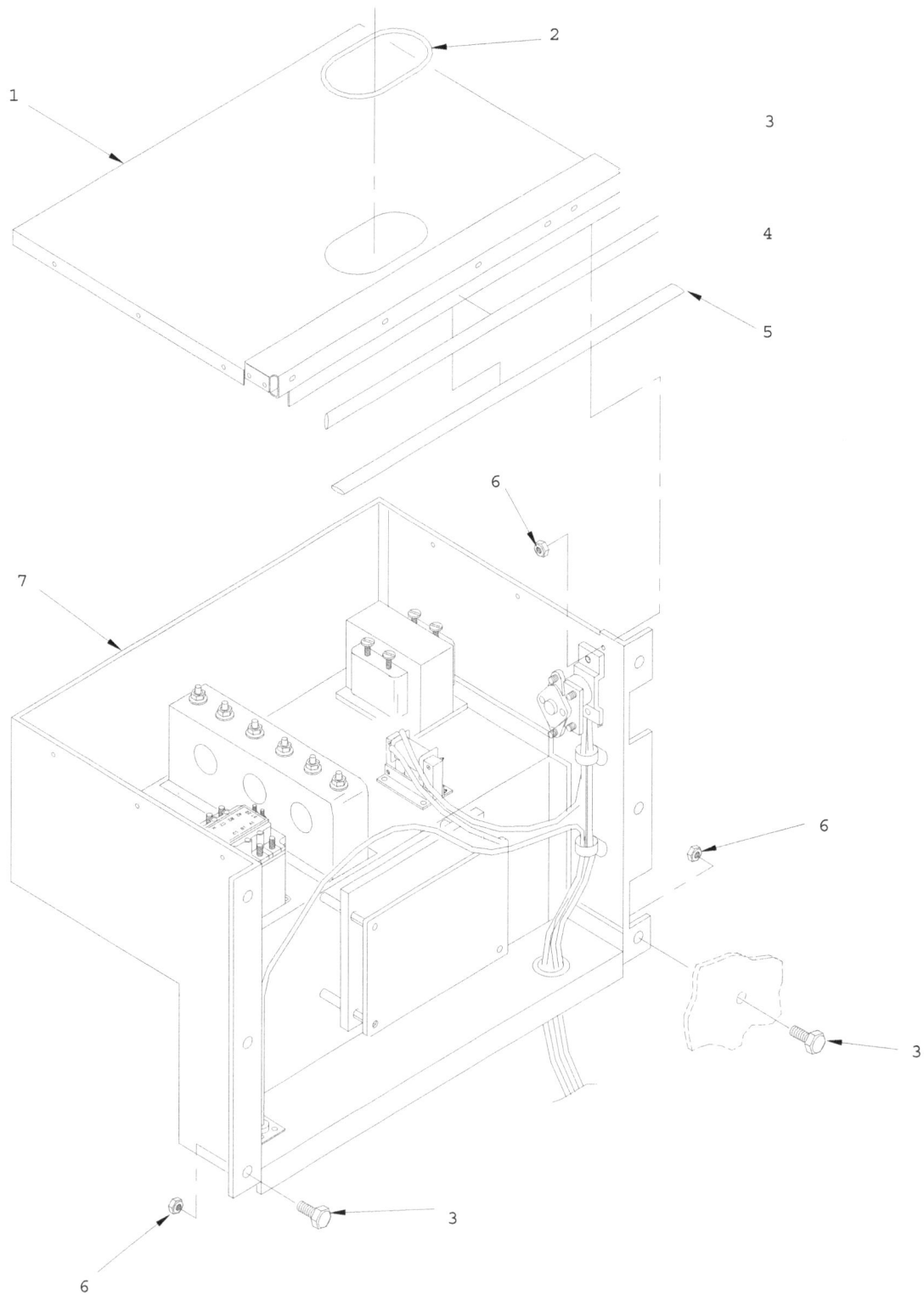

Figure 14. Output Box Assembly (Sheet 1 of 3)

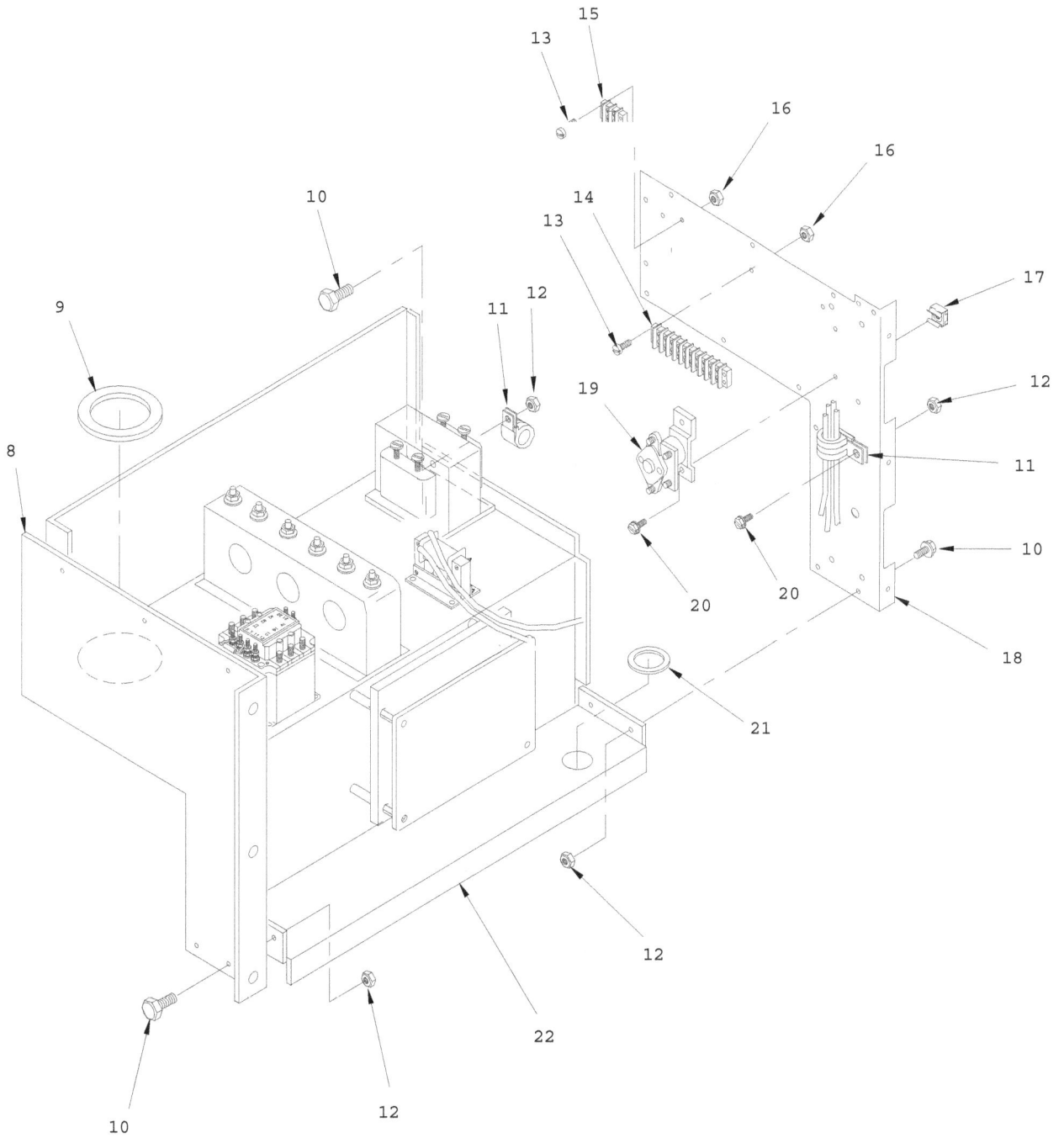

Figure 14. Output Box Assembly (Sheet 2 of 3)

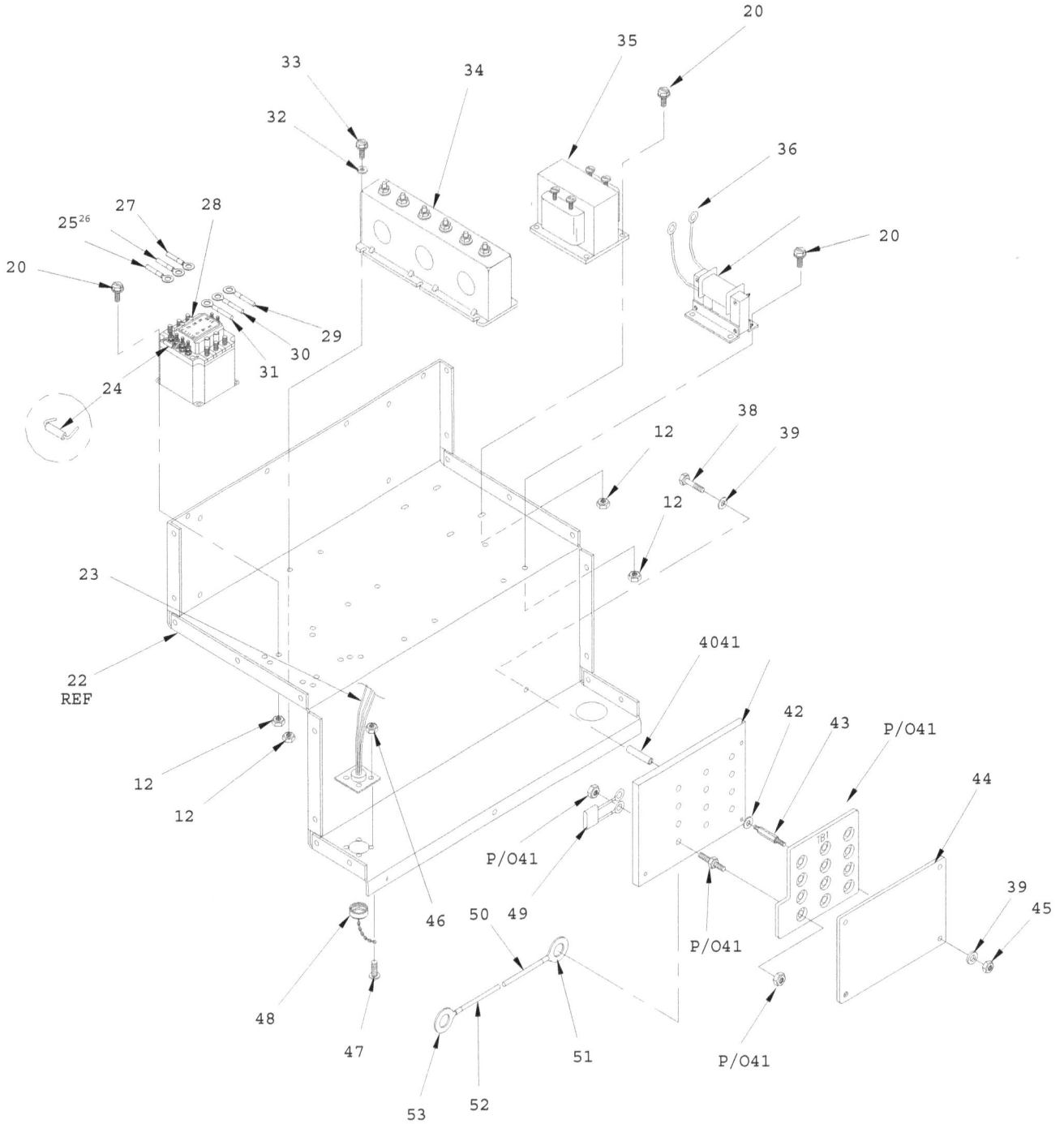

Figure 14. Output Box Assembly (Sheet 3 of 3)

(1)		(2)		(3)	(4)	(5)	(6)
					SMR CODE		
ITEM		AIR			PART	DESCRIPTION AND	
NO	ARMY	FORCE	USMC	CAGEC	NUMBER	USABLE ON CODE (UOC)	QTY

GROUP 07 OUTPUT BOX ASSEMBLY
FIGURE 14 OUTPUT BOX ASSEMBLY

1	XBFZZ	XB	XBFZZ	30554	88-22701	TOP,OUTPUT BOX........................1	2
	MFFZZ	M000	MFFZZ	30554	88-22704	SEAL,NONMETALLIC SP....................V	
						MAKE FROM P/N ZX-4134 (1X968), CUT	
						AS REQ	
3	PAFZZ	PA000	PAFZZ	19207	12325869	BOLT,MACHINE........................19	
4	PAOZZ	PA000	PAOZZ	77824	5798-190Z-3-75.0	SEAL,NONMETALLIC SP....................1	5
	MOOZZ	M000	M0000	30554	88-22705	GASKET................................V	
						MAKE FROM P/N 20941 (56329), CUT AS REQ	
6	PAOZZ	PA000	PAOZZ	19617	HW310	NUT,PLAIN,ASSEMBLED...................19	
7	XBFFF	XB	XBFFF	30554	96-23600	OUTPUT BOX ASSEMBLY...................1	
8	XBFZZ	XB	XBFZZ	30554	88-21932	PANEL,OUTPUT BOX,LS...................1	
9	XBFZZ	XB	XBFZZ	76385	Z-3170	GROMMET,NONMETALLIC...................1	
10	PAFZZ	PA000	PAFZZ	19207	12325869	BOLT,MACHINE.........................9	
11	PAFZZ	PA000	PAFZZ	22175	JM75LC6SS14R	CLAMP,LOOP.............................2	
12	PAOZZ	PA000	PAOZZ	19617	HW310	NUT,PLAIN,ASSEMBLED...................25	
13	PAFZZ	PA000	PAFZZ	30554	69-662-36	SCREW,ASSEMBLED WAS...................4	
14	PAFZZ	PA000	PAFZZ	9R803	3300-10-XP-74	MARKER STRIP,TERMIN...................1	
15	PAFZZ	PA000	PAFZZ	30554	88-21182-2	MARKER STRIP,TERMIN...................1	
16	PAFZZ	PA000	PAFZZ	78189	511-081800-00	NUT,PLAIN,ASSEMBLED...................4	
17	PAFZZ	PA000	PAFZZ	78553	C7931-1032-38	NUT,PLAIN,CLINCH.....................1	
18	XBFZZ	XB	XBFZZ	30554	96-23678	PANEL,OUTPUT BOX,RS...................1	
19	PAFZZ	PA000	PAFZZ	81640	SM5008	RELAY,ELECTROMAGNET...................1	
20	PAFZZ	PA000	PAFZZ	45722	P-15121-64	SCREW,ASSEMBLED WAS...................16	
21	XBFZZ	XB	XBFZZ	96906	MS35489-27	GROMMET,NONMETALLIC...................1	
22	XBFZZ	XB	XBFZZ	30554	88-22039	PANEL,OUTPUT BOX,RE...................1	
23	XBFFF	XB	XBFFF	30554	96-23575	WIRING HARNESS,OUTP...................1	
						(SEE FIGURE 15 FOR PARTS BREAKDOWN)	
24	PAFZZ	PA000	PAFZZ	30554	88-22418-2	SEMICONDUCTOR DEVIC...................2	
25	AFFFF	M000	MFFFF	30554	88-22126-7	CABLE ASSEMBLY,AC P...................1	
						(SEE FIGURE 16 FOR PARTS BREAKDOWN)	
26	AFFFF	M000	MFFFF	30554	88-22126-6	CABLE ASSEMBLY,AC P...................1	
						(SEE FIGURE 16 FOR PARTS BREAKDOWN)	
27	AFFFF	M000	MFFFF	30554	88-22126-5	CABLE ASSEMBLY,AC P...................1	
						(SEE FIGURE 16 FOR PARTS BREAKDOWN)	
28	PAFZZ	PA000	PAFZZ	81640	9565H150	CONTACTOR,MAGNETIC...................1	
29	AFFFF	M000	MFFFF	30554	88-22126-1	CABLE ASSEMBLY,AC P...................1	
						(SEE FIGURE 16 FOR PARTS BREAKDOWN)	
30	AFFFF	M000	MFFFF	30554	88-22126-2	CABLE ASSEMBLY,AC P...................1	
						(SEE FIGURE 16 FOR PARTS BREAKDOWN)	
31	AFFFF	M000	MFFFF	30554	88-22126-3	CABLE ASSEMBLY,AC P...................1	
						(SEE FIGURE 16 FOR PARTS BREAKDOWN)	
32	PAFZZ	PA000	PAFZZ	96906	MS51412-2	WASHER,FLAT..........................6	

(1)	(2)		(3)	(4)	(5)	(6)
	AIR			SMR CODE	DESCRIPTION AND	
ITEM				PART		
NO	ARMY	FORCE	USMC	CAGEC	NUMBER	USABLE ON CODE (UOC)	QTY

(1) ITEM NO	(2) ARMY	AIR FORCE	USMC	(3) CAGEC	(4) PART NUMBER SMR CODE	(5) DESCRIPTION AND USABLE ON CODE (UOC)	(6) QTY
33	PAFZZ	PA000	PAFZZ	78189	61-101041-90-014	SCREW,ASSEMBLED WAS...................6	
34	PAFZZ	PA000	PAFZZ	60177	19390	TRANSFORMER,CURRENT...................1	
35	PAFZZ	PA000	PAFZZ	14407	CC1-00380-110L	TRANSFORMER,DISCRIM...................1	
36	PAFZZ	PA000	PAFZZ	06383	PN18-6LF-C	TERMINAL,LUG...........................2	
37	PAFZZ	PA000	PAFZZ	30554	88-21015	TRANSFORMER,POWER.....................1	
38	PAFZZ	PA000	PAFZZ	80204	B1821BH025C275N	SCREW,CAP,HEXAGON H...................4	
39	PAFZZ	PA000	PAFZZ	96906	MS27183-10	WASHER,FLAT...........................12	
40	XB0ZZ	XB	XB0ZZ	30554	88-21669	MOUNT,BOARD,RECONNE...................4	
41	PAFZZ	PA000	PAFZZ	30554	88-22437	TERMINAL BOARD........................1	
42	PAFZZ	PA000	PAFZZ	97403	13230E6744-63	WASHER,FLAT...........................4	
43	XBFZZ	XB	XBFZZ	30554	72-2098-2	POST,ELECTRICAL-MEC...................4	
44	PAFZZ	PA000	PAFZZ	30554	88-22509	COVER,DISTRIBUTION....................1	
45	PAFZZ	PA000	PAFZZ	30554	88-20568-1	NUT,PLAIN,CASTELLAT...................4	
46	PAFZZ	PA000	PAFZZ	78189	501-040800-00	NUT,PLAIN,ASSEMBLED...................4	
47	PAFZZ	PA000	PAFZZ	45722	69-662-5	SCREW,ASSEMBLED WAS...................4	
48	PAFZZ	PA000	PAFZZ	77820	MS25043-18DA	COVER,ELECTRICAL,CO...................1	
49	PAFZZ	PA000	PAFZZ	30554	88-22758	CAPACITOR ASSEMBLY....................3	
50	AFFFF	M000	MFFFF	30554	88-22126-4	CABLE ASSEMBLY,POWE...................1	
51	PAFZZ	PA000	PAFZZ	56501	RG9731	TERMINAL,LUG..........................1	52
52	MFFZZ	M000	MFFZZ	30554	88-20540-1	WIRE,ELECTRIC.........................V	
						MAKE FROM P/N MD66530GNB (90484)	
53	PAFZZ	PA000	PAFZZ	96906	MS25036-128	TERMINAL,LUG..........................1	

END OF FIGURE

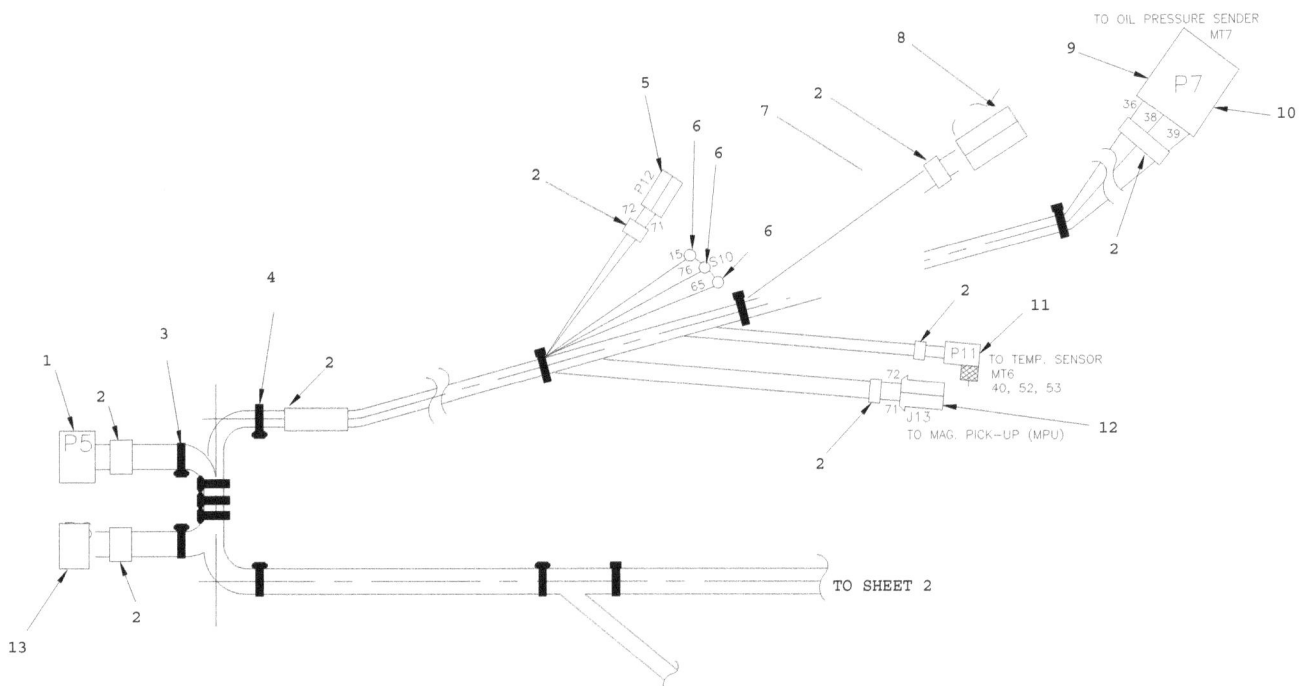

Figure 15. Output Box Wiring Harness (Sheet 1 of 3)

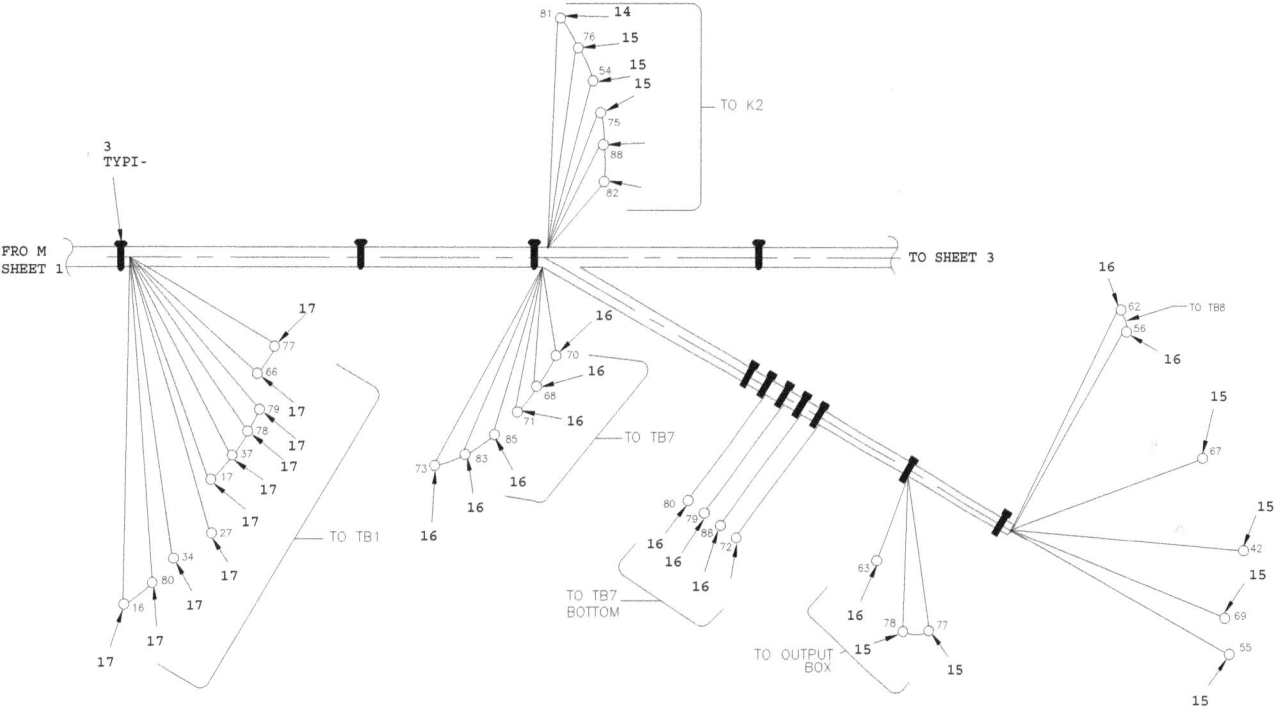

Figure 15. Output Box Wiring Harness (Sheet 2 of 3)

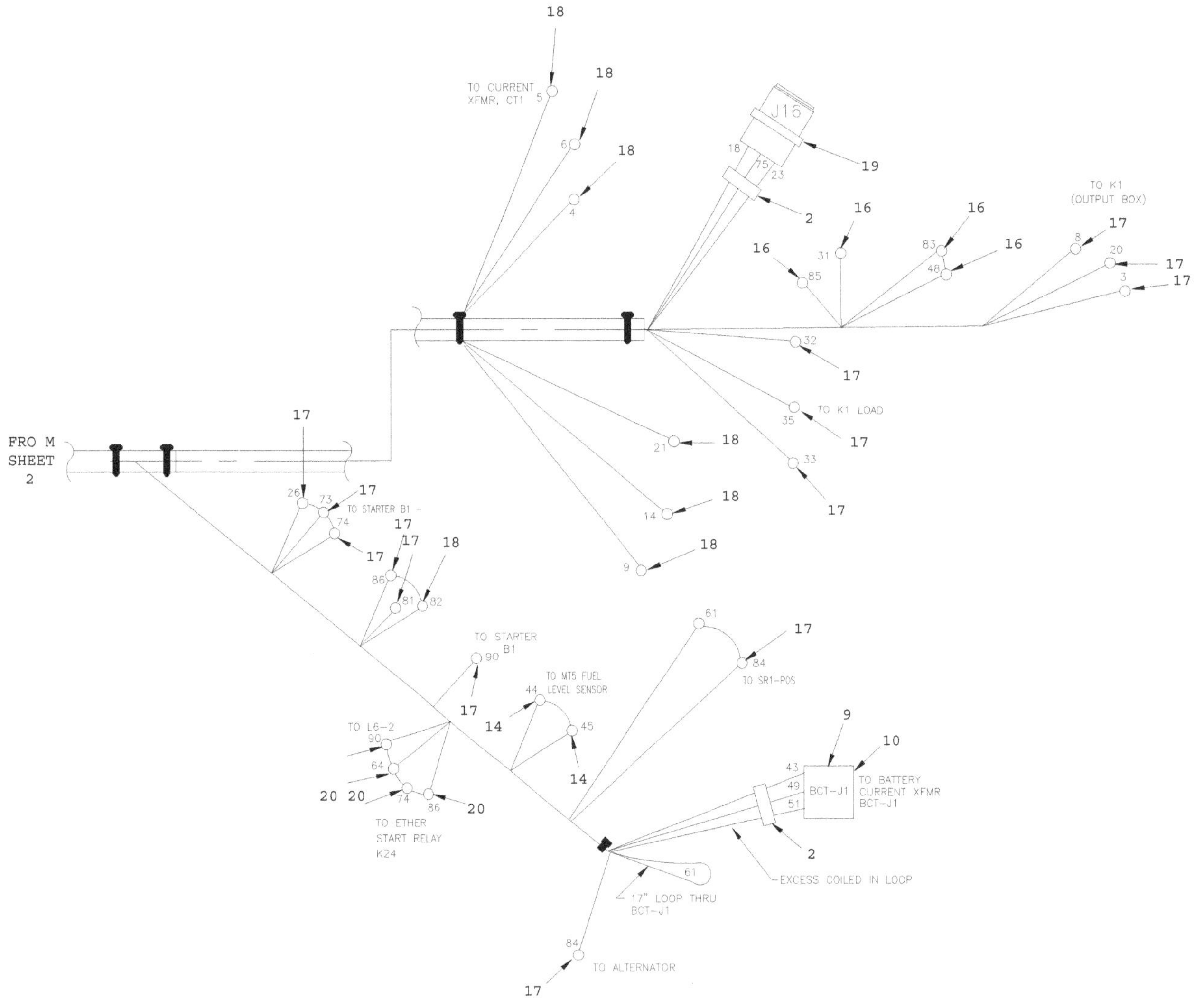

Figure 15. Output Box Wiring Harness (Sheet 3 of 3)

(1)		(2)		(3)	(4)	(5)	(6)
					SMR CODE		
ITEM		AIR			PART	DESCRIPTION AND	
NO	ARMY	FORCE	USMC	CAGEC	NUMBER	USABLE ON CODE (UOC)	QTY

GROUP 0702 OUTPUT BOX WIRING HARNESS
FIGURE 15 OUTPUT BOX WIRING HARNESS

1	PAFZZ	PA000	PAFZZ	77820	97-3108B-28-15P	CONNECTOR,PLUG........................1	
2	XBFZZ	XB	XBFZZ	06383	PLM2S	STRAP,IDENTIFICATIO....................8	
3	PAFZZ	PA000	PAFZZ	06383	PLT2S	STRAP,TIEDOWN,ELECT...................V	
4	PAFZZ	PA000	PAFZZ	96906	MS3367-5-9	STRAP,TIEDOWN,ELECT...................V	
5	PAFZZ	PA000	PAFZZ	27264	03-09-1022	CONNECTOR,BODY,RECE...................1	
6	PAFZZ	PA000	PAFZZ	98410	BB-837-06	TERMINAL,LUG..........................3	7
	PAFZZ	PA000	PAFZZ	8W764	8719	CABLE,SPECIAL PURPO...................V	
8	PAFZZ	PA000	PAFZZ	1JB11	15300027	CONNECTOR,SNAP........................1	
9	PAFZZ	PA000	PAFZZ	27264	03-09-1042	CONNECTOR BODY,PLUG...................2	
10	PAFZZ	PA000	PAFZZ	27264	02-09-1104	CONTACT,ELECTRICAL....................8	
11	PAFZZ	PA000	PAFZZ	30554	96-23546	CABLE ASSEMBLY,POWE...................1	
12	PAFZZ	PA000	PAFZZ	30554	96-23680-1	CONNECTOR,MODIFIED....................1	
13	PAFZZ	PA000	PAFZZ	77820	97-3108B-28-15S	CONNECTOR,PLUG........................1	
14	PAFZZ	PA000	PAFZZ	98410	BB-837-10	TERMINAL,LUG.........................32	
15	PAFZZ	PA000	PAFZZ	98410	BB-837-08	TERMINAL,LUG..........................4	
16	PAFZZ	PA000	PAFZZ	98410	BB-8707-06	TERMINAL,SPADE.......................21	
17	PAFZZ	PA000	PAFZZ	98410	BB-818-38	TERMINAL,LUG.........................20	
18	PAFZZ	PA000	PAFZZ	98410	BB-8194-08	TERMINAL,SPADE........................7	
19	PAFZZ	PA000	PAFZZ	96906	MS3102R18-11P	CONNECTOR,RECEPTACL...................1	
20	PAFZZ	PA000	PAFZZ	15912	RB2573	TERMINAL,QUICK DISC...................3	
21	PAFZZ	PA000	PAFZZ	56501	RA2573M	TERMINAL,DISCONNECT...................2	22
	PAFZZ	PA000	PAFZZ	30554	88-20450-4	WIRE,ELECTRICAL.......................V	

END OF FIGURE

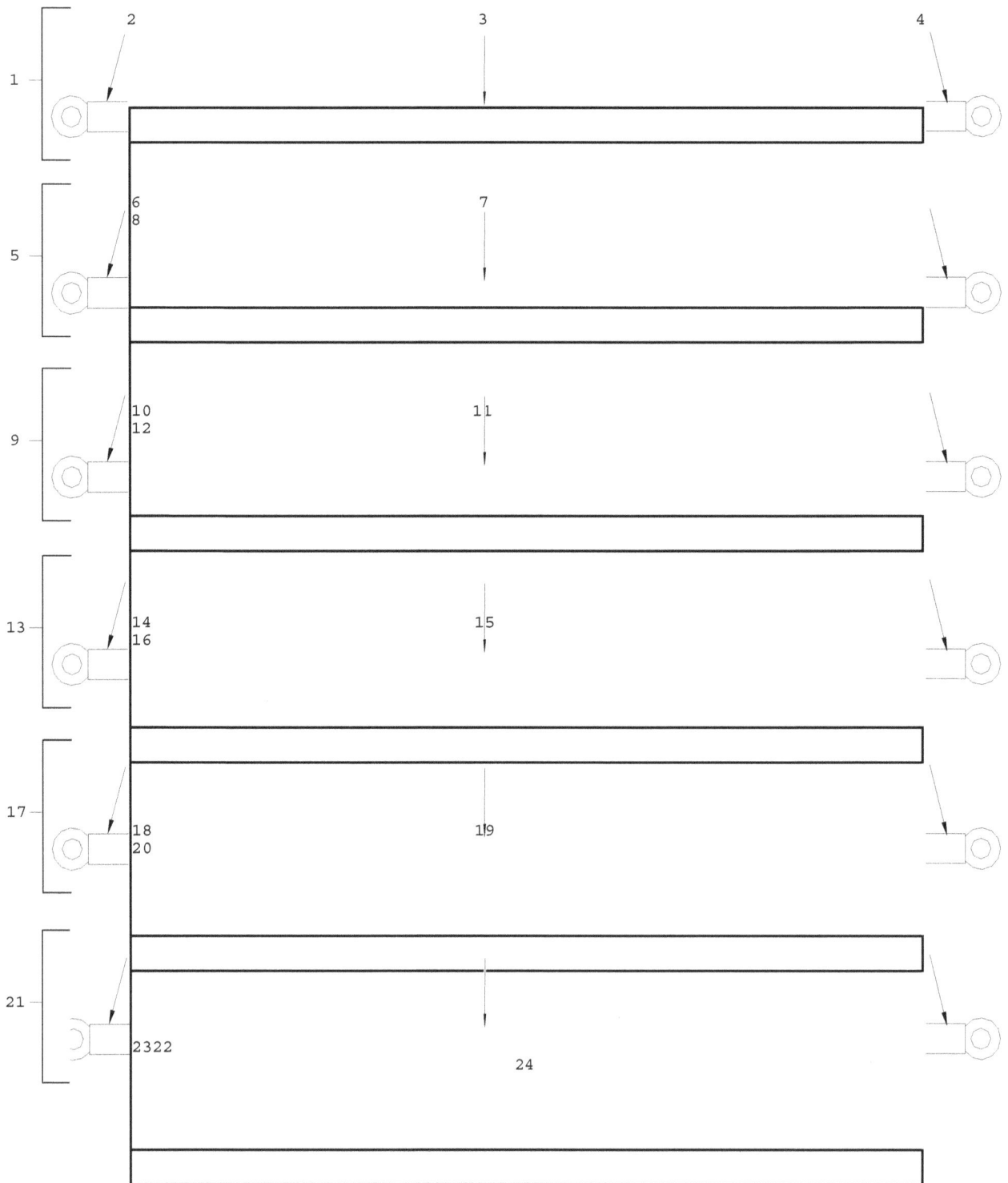

Figure 16. AC Power Cables

(1)	(2)			(3)	(4)	(5)	(6)
					SMR CODE		
ITEM		AIR			PART	DESCRIPTION AND	
NO	ARMY	FORCE	USMC	CAGEC	NUMBER	USABLE ON CODE (UOC)	QTY

GROUP 0708 POWER CABLES
FIGURE 16 POWER CABLES

1	AFFFF	M000	MFFFF	30554	88-22126-1	CABLE ASSEMBLY,AC P...................1	
2	PAFZZ	PA000	PAFZZ	98410	G-775-14	TERMINAL,LUG..........................1	3
MFFZZ	M000	MFFZZ		30554	88-20540-11	WIRE,ELECTRIC.........................V	
						MAKE FROM P/N MD66530GNB (90484),	
						14.5 IN	
4	PAFZZ	PA000	PAFZZ	56501	RG9731	TERMINAL,LUG..........................1	
5	AFFFF	M000	MFFFF	30554	88-22126-1	CABLE ASSEMBLY,AC P...................1	
6	PAFZZ	PA000	PAFZZ	98410	G-775-14	TERMINAL,LUG..........................1	7
MFFZZ	M000	MFFZZ		30554	88-20540-11	WIRE,ELECTRIC.........................V	
						MAKE FROM P/N MD66530GNB (90484),	
						17.0 IN	
8	PAFZZ	PA000	PAFZZ	56501	RG9731	TERMINAL,LUG..........................1	
9	AFFFF	M000	MFFFF	30554	88-22126-3	CABLE ASSEMBLY,AC P...................1	
10	PAFZZ	PA000	PAFZZ	98410	G-775-14	TERMINAL,LUG..........................1	11
MFFZZ	M000	MFFZZ		30554	88-20540-11	WIRE,ELECTRIC.........................V	
						MAKE FROM P/N MD66530GNB (90484),	
						20.0 IN	
12	PAFZZ	PA000	PAFZZ	56501	RG9731	TERMINAL,LUG..........................1	
13	AFFFF	M000	MFFFF	30554	88-22126-5	CABLE ASSEMBLY,AC P...................1	
14	PAFZZ	PA000	PAFZZ	98410	G-775-14	TERMINAL,LUG..........................1	15
MFFZZ	M000	MFFZZ		30554	88-20540-11	WIRE,ELECTRIC.........................V	
						MAKE FROM P/N MD66530GNB (90484),	
						31.0 IN	
16	PAFZZ	PA000	PAFZZ	96906	MS25036-128	TERMINAL,LUG..........................1	
17	AFFFF	M000	MFFFF	30554	88-22126-6	CABLE ASSEMBLY,AC P...................1	
18	PAFZZ	PA000	PAFZZ	98410	G-775-14	TERMINAL,LUG..........................1	19
MFFZZ	M000	MFFZZ		30554	88-20540-11	WIRE,ELECTRIC.........................V	
						MAKE FROM P/N MD66530GNB (90484),	
						33.0 IN	
20	PAFZZ	PA000	PAFZZ	96906	MS25036-128	TERMINAL,LUG 1	
21	AFFFF	M000	MFFFF	30554	88-22126-7	CABLE ASSEMBLY,AC P...................1	
22	PAFZZ	PA000	PAFZZ	98410	G-775-14	TERMINAL,LUG..........................1	23
MFFZZ	M000	MFFZZ		30554	88-20540-11	WIRE,ELECTRIC.........................V	
						MAKE FROM P/N MD66530GNB (90484),	
						35.0 IN	
24	PAFZZ	PA000	PAFZZ	96906	MS25036-128	TERMINAL,LUG........................11	

END OF FIGURE

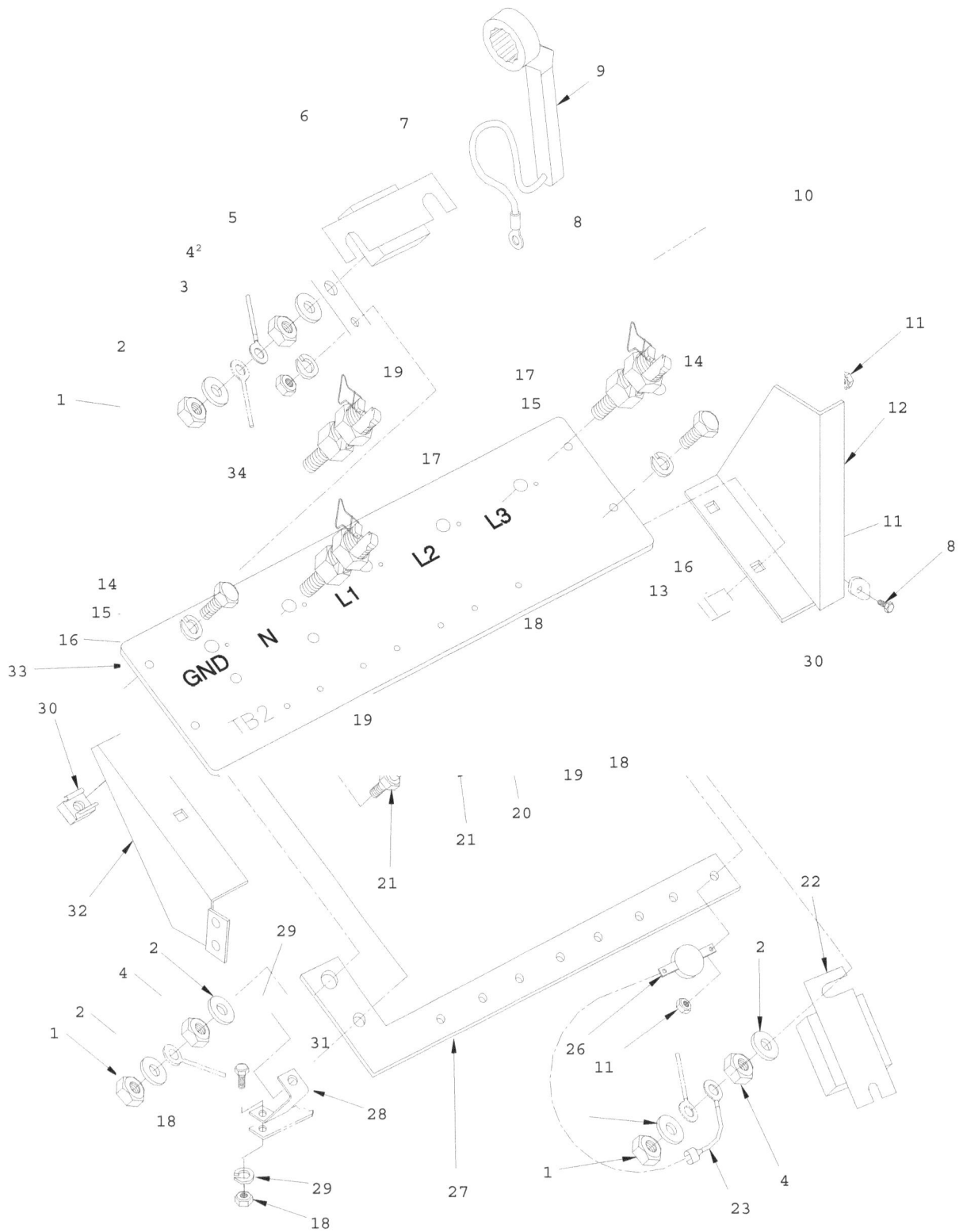

Figure 17. Load Output Terminal Board Assembly

(1)				(3)	(4)	(5)	(6)
		(2)			SMR CODE		
ITEM		AIR			PART	DESCRIPTION AND	
NO	ARMY	FORCE	USMC	CAGEC	NUMBER	USABLE ON CODE (UOC)	QTY

GROUP 0707 LOAD OUTPUT TERMINAL BOARD
FIGURE 17 LOAD OUTPUT TERMINAL BOARD

1	PAOZZ	PAO00	PAOZZ	08928	21NTE813	NUT,SELF-LOCKING,HE...................5	
2	PAOZZ	PAO00	PAOZZ	30554	88-20564-15	WASHER,FLAT.........................10	
3	AO000	MO00	MO000	30554	88-20305-5	WIRE,VARISTOR........................1	
						(REFER TO FIG. 18 FOR BREAKDOWN)	
4	PAOZZ	PAO00	PAOZZ	63857	BS4-1801PC15	NUT,PLAIN,HEXAGON....................5	
5	XBOZZ	XB	XBOZZ	30554	88-22163-1	BAR,BUS,GROUND.......................1	
6	PAOZZ	PAO00	PAOZZ	30554	88-22762	FILTER ASSEMBLY,ELE..................1	
7	PAOZZ	PAO00	PAOZZ	30554	88-22469	CORD ASSEMBLY,FIBRO..................1	
8	PAOZZ	PAO00	PAOZZ	19207	12325869	BOLT,MACHINE.........................3	
9	PAOZZ	PAO00	PAOZZ	30554	88-21147	WRENCH,BOX...........................1	
10	XBOZZ	XB	XBOZZ	30554	88-22136	BRACKET,WRENCH.......................1	
11	PAOZZ	PAO00	PAOZZ	19617	HW310	NUT,PLAIN,ASSEMBLED.................13	
12	XBOZZ	XB	XBOZZ	30554	88-21933	SUPPORT,LOAD BOARD...................1	
13	PAOZZ	PAO00	PAOZZ	30554	88-20260-25	SCREW,CAP,HEXAGON H..................8	
14	PAOZZ	PAO00	PAOZZ	80204	B1821BH025C125N	SCREW,CAP,HEXAGON H..................4	
15	PAOZZ	PAO00	PAOZZ	97403	13230E6744-63	WASHER,FLAT..........................4	
16	PAOZZ	PAO00	PAOZZ	96906	MS27183-10	WASHER,FLAT..........................4	
17	PAOZZ	PAO00	PAOZZ	96906	MS39347-5	TERMINAL,STUD........................5	
18	PAOZZ	PAO00	PAOZZ	30554	88-22786	NUT,SELF-LOCKING,HE..................4	
19	PAOZZ	PAO00	PAOZZ	30554	88-20564-14	WASHER,FLAT..........................2	
20	XBOZZ	XB	XBOZZ	30554	88-22146	BAR,BUS..............................1	
21	PAOZZ	PAO00	PAOZZ	30554	72-2236	TERMINAL,STUD........................2	
22	PAOZZ	PAO00	PAOZZ	30554	88-22761	FILTER ASSEMBLY,ELE..................3	
23	AO000	MO00	MO000	30554	88-20305-3	WIRE,VARISTOR........................1	
						(SEE FIGURE 18 FOR PARTS BREAKDOWN)	
24	AO000	MO00	MO000	30554	88-20305-1	WIRE,VARISTOR........................1	
						(SEE FIGURE 18 FOR PARTS BREAKDOWN)	
25	AO000	MO00	MO000	30554	88-20305-2	WIRE,VARISTOR........................1	
						(SEE FIGURE 18 FOR PARTS BREAKDOWN)	
26	PAOZZ	PAO00	PAOZZ	81483	40-1389	RESISTOR,VOLTAGE SE..................4	
27	XBOZZ	XB	XBOZZ	30554	88-21775	BAR,GROUND PLANE.....................1	
28	XBOZZ	XB	XBOZZ	30554	88-22318	STRAP,GROUND.........................1	
29	PAOZZ	PAO00	PAOZZ	97403	13230E6744-65	WASHER,LOCK..........................3	
30	PAOZZ	PAO00	PAOZZ	78553	C7988-1420-38	NUT,PLAIN,CLINCH.....................4	
31	PAOZZ	PAO00	PAOZZ	80204	B1821BH038C100N	SCREW,CAP,HEXAGON H..................1	
32	XBOZZ	XB	XBOZZ	30554	88-21767	SUPPORT,LOAD BOARD...................1	
33	XBOZZ	XB	XBOZZ	30554	88-21667	BOARD,LOAD TERMINAL..................1	
34	PAOZZ	PAO00	PAOZZ	08928	21NTE616	NUT,SELF-LOCKING,HE..................1	

END OF FIGURE

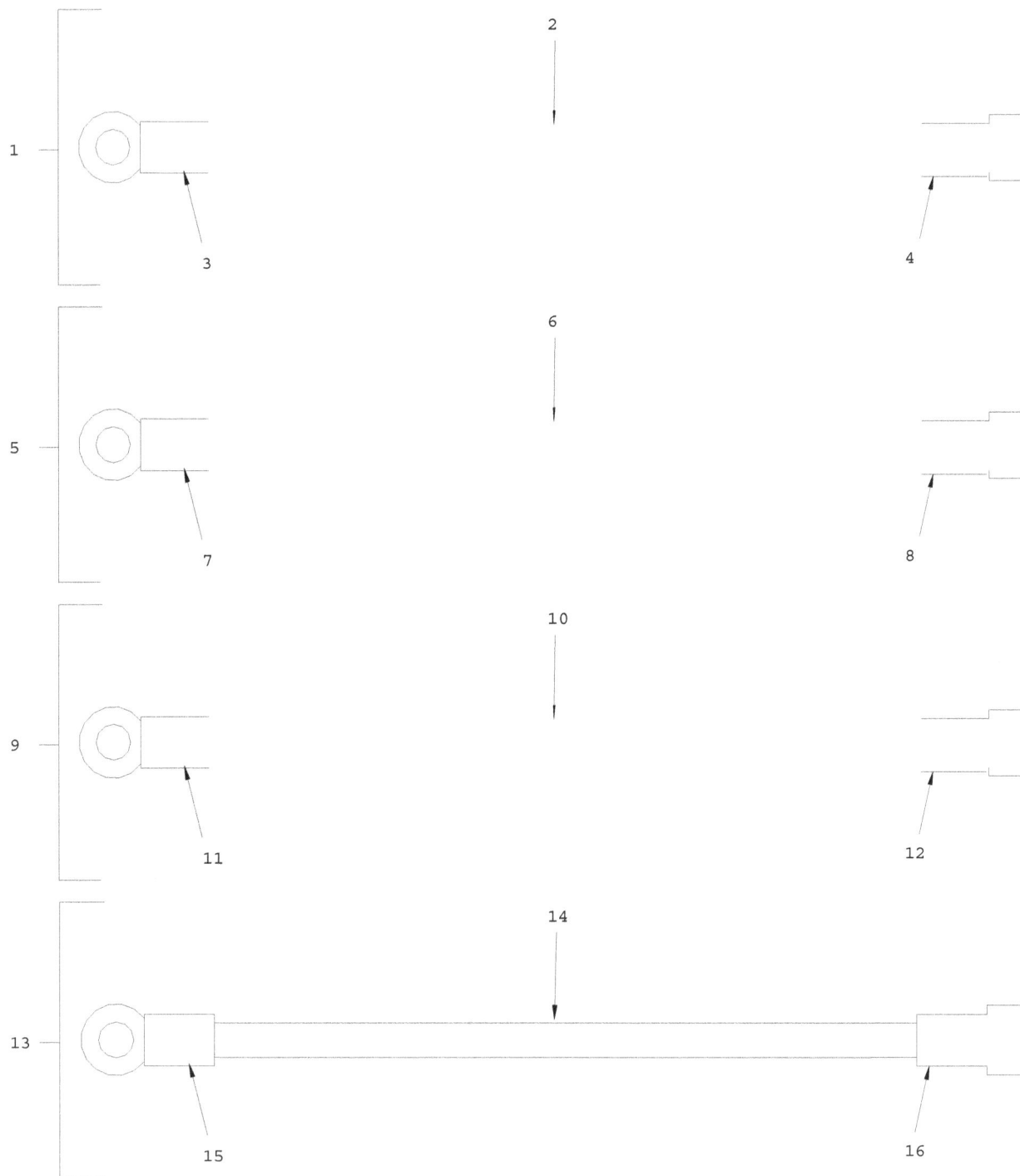

Figure 18.   Varistor Wires

(1)		(2)		(3)	(4)	(5)	(6)
					SMR CODE		
ITEM		AIR			PART	DESCRIPTION AND	
NO	ARMY	FORCE	USMC	CAGEC	NUMBER	USABLE ON CODE (UOC)	QTY

GROUP 0709 VARISTOR WIRES
FIGURE 18 VARISTOR WIRES

1	A0000	M000	M0000	30554	88-20305-1	WIRE,VARISTOR.........................1
2	M00ZZ	M000	M0000	30554	88-20450-4	WIRE,ELECTRICAL.......................1

MAKE FROM P/N M5086/2-16-9 (81349),
AS REQ

3	PA0ZZ	PA000	PA0ZZ	15912	RB2573	TERMINAL,QUICK DISC...................1
4	PA0ZZ	PA000	PA0ZZ	98410	B-801-12HDX	TERMINAL,LUG..........................1
5	A0000	M000	M0000	30554	88-20305-2	WIRE,VARISTOR.........................1
6	M00ZZ	M000	M0000	30554	88-20450-4	WIRE,ELECTRICAL.......................1

MAKE FROM P/N M5086/2-16-9 (81349),
AS REQ

7	PA0ZZ	PA000	PA0ZZ	15912	RB2573	TERMINAL,QUICK DISC...................1
8	PA0ZZ	PA000	PA0ZZ	98410	B-801-12HDX	TERMINAL,LUG..........................1
9	A0000	M000	M0000	30554	88-20305-3	WIRE,VARISTOR 1
10	M00ZZ	M000	M0000	30554	88-20450-4	WIRE,ELECTRICAL.......................1

MAKE FROM P/N M5086/2-16-9 (81349),
AS REQ

11	PA0ZZ	PA000	PA0ZZ	15912	RB2573	TERMINAL,QUICK DISC...................1
12	PA0ZZ	PA000	PA0ZZ	98410	B-801-12HDX	TERMINAL,LUG..........................1
13	A0000	M000	M0000	30554	88-20305-5	RESISTOR,VOLTAGE SE...................1
14	M00ZZ	M000	M0000	30554	88-20450-4	WIRE,ELECTRICAL.......................1

MAKE FROM P/N M5086/2-16-9 (81349),
AS REQ

15	PA0ZZ	PA000	PA0ZZ	15912	RB2573	TERMINAL,QUICK DISC...................1
16	PA0ZZ	PA000	PA0ZZ	98410	B-801-12HDX	TERMINAL,LUG..........................1

END OF FIGURE

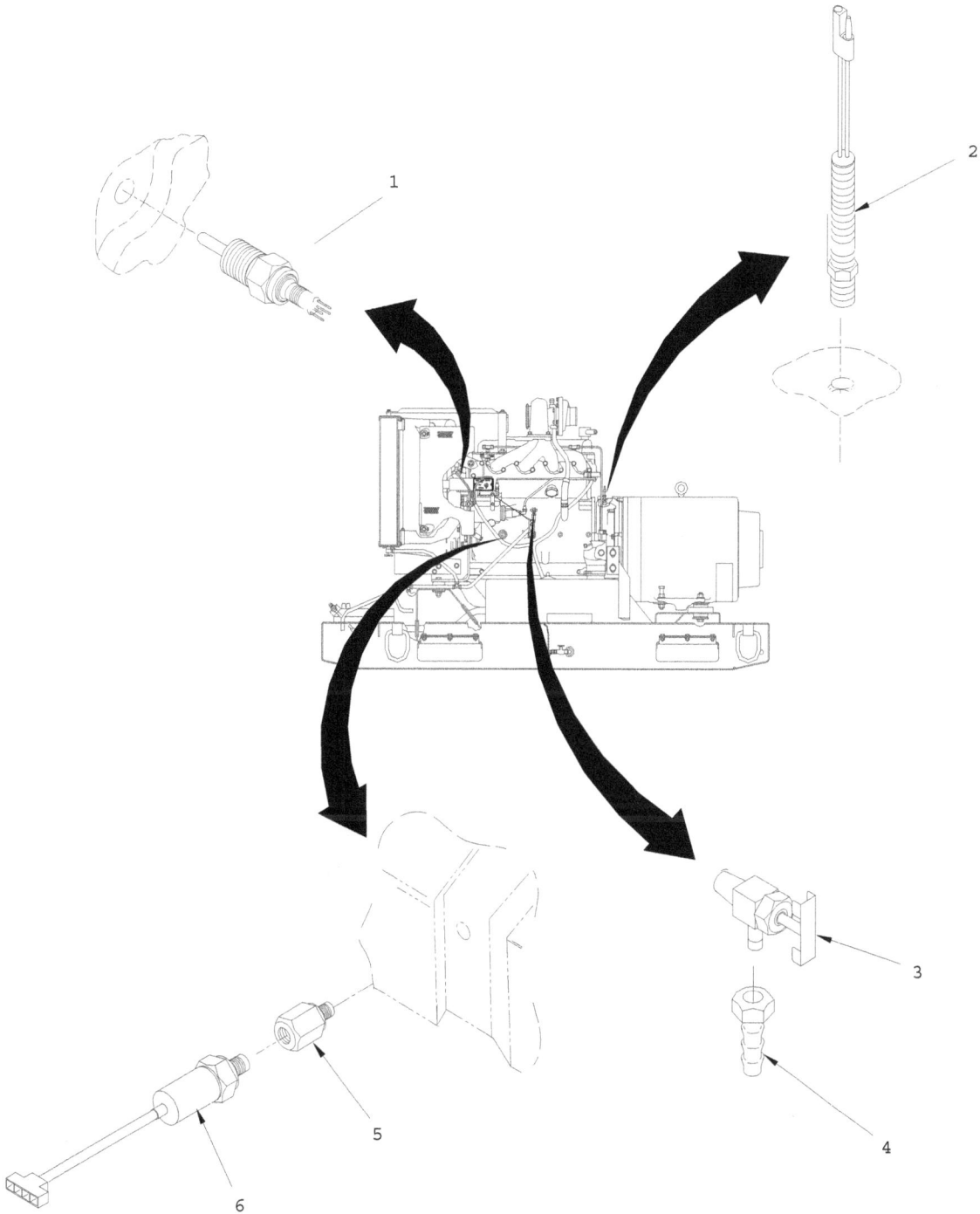

Figure 19. Engine Accessories (Sheet 1 of 4)

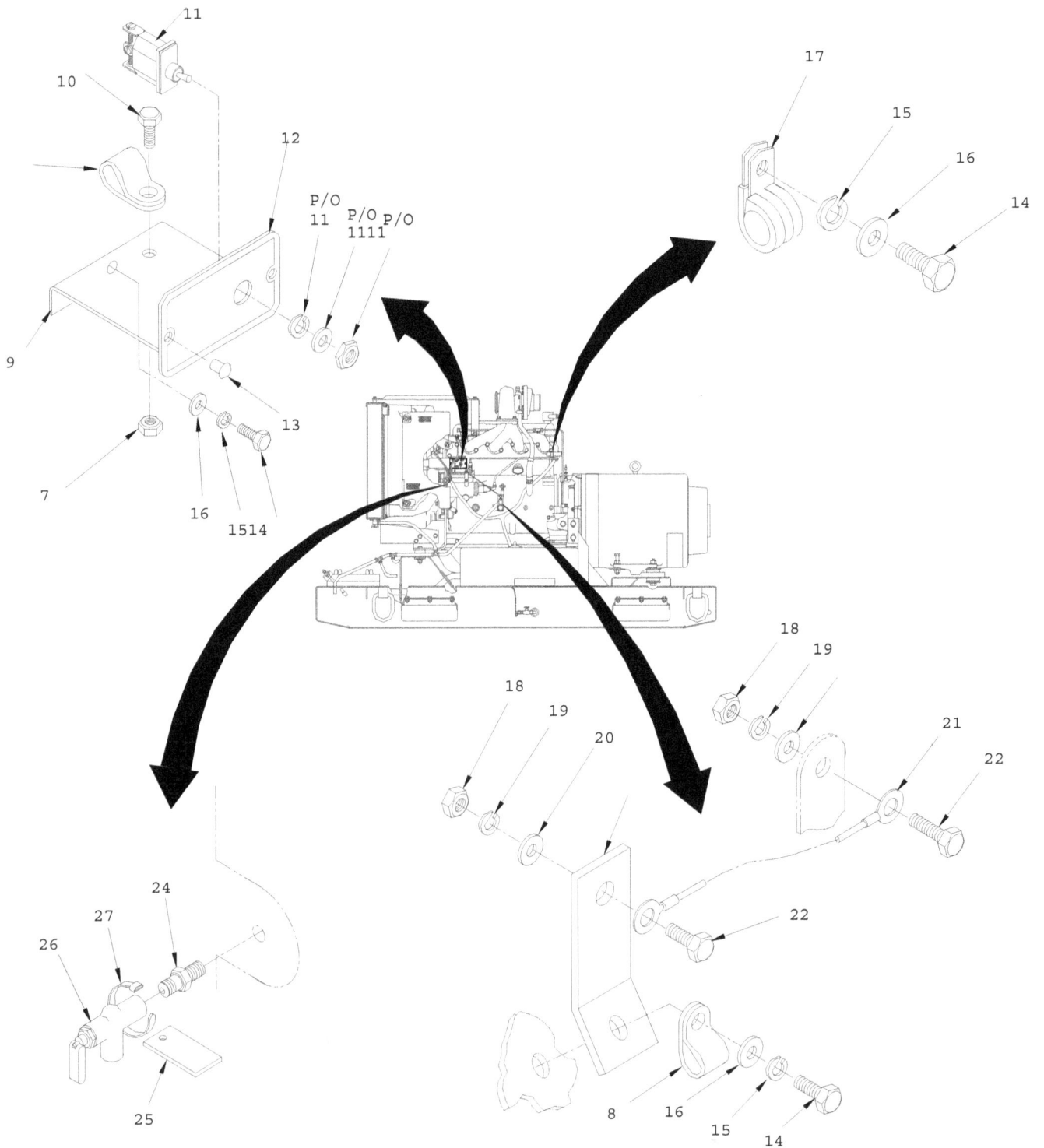

Figure 19. Engine Accessories (Sheet 2 of 4)

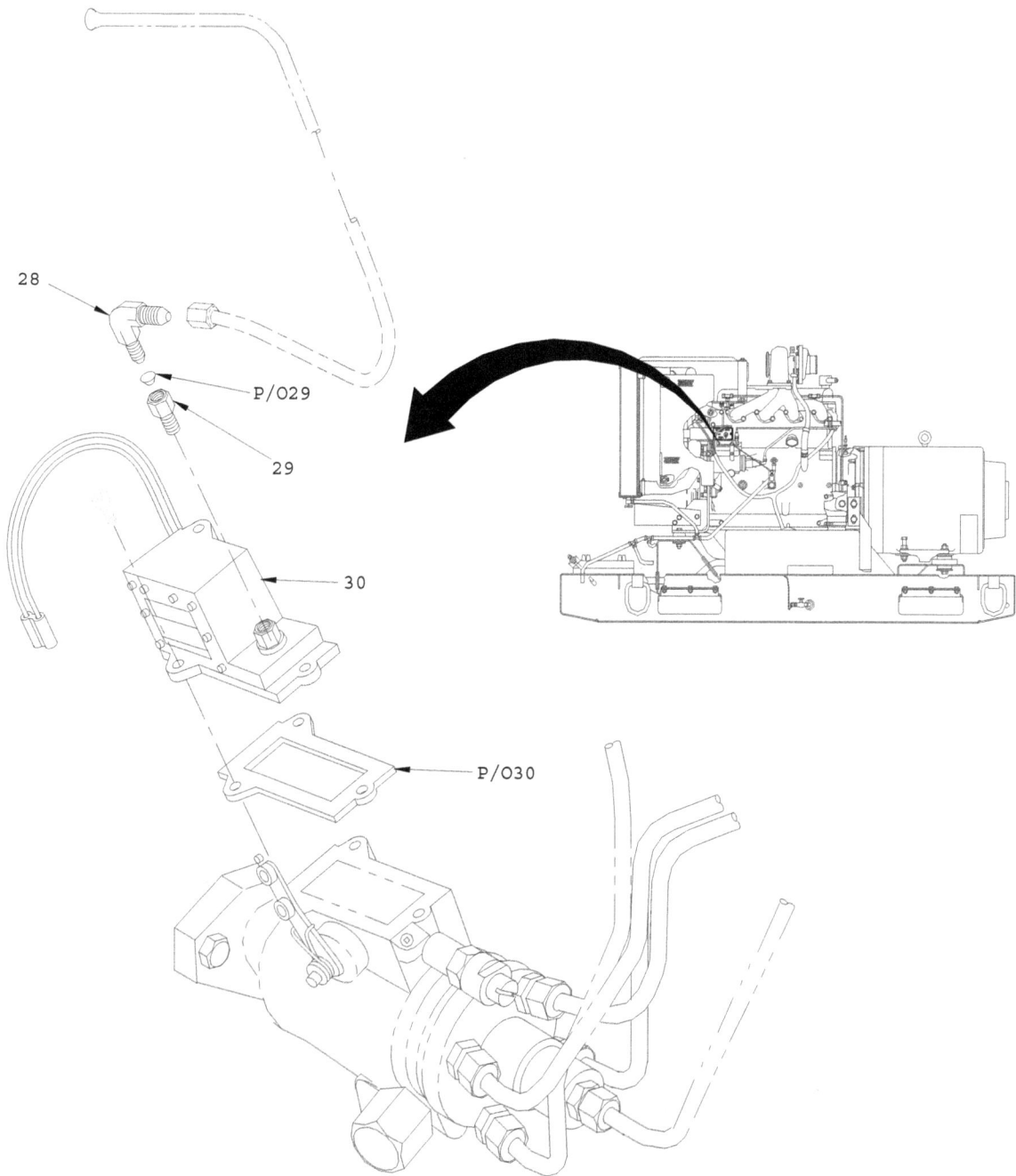

Figure 19. Engine Accessories (Sheet 3 of 4)

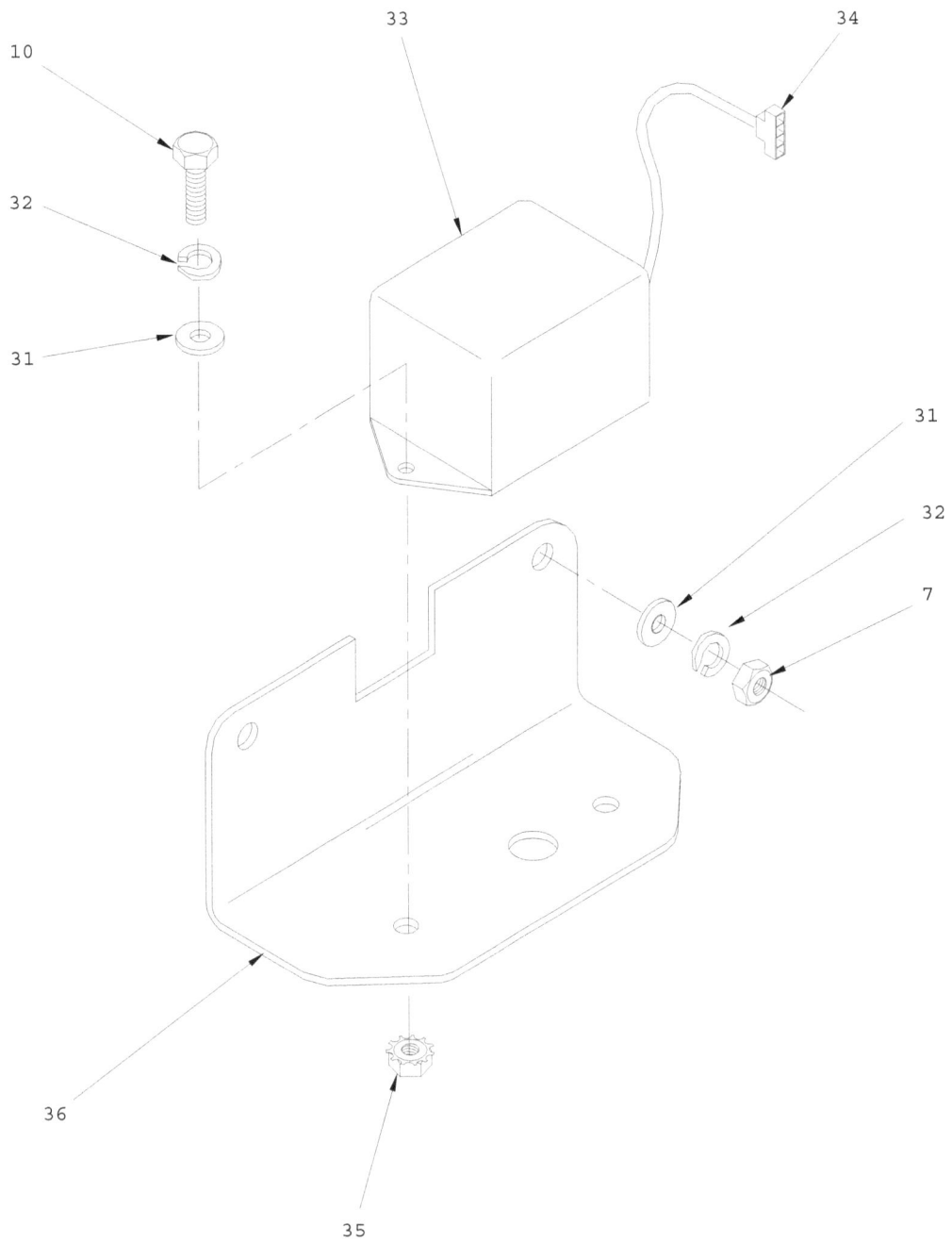

Figure 19. Engine Accessories (Sheet 4 of 4)

(1)	(2)			(3)	(4)	(5)	(6)
					SMR CODE		
ITEM	AIR				PART	DESCRIPTION AND	
NO	ARMY	FORCE	USMC	CAGEC	NUMBER	USABLE ON CODE (UOC)	QTY

GROUP 08 ENGINE ACCESSORIES
FIGURE 19 ENGINE ACCESSORIES

1	PAOZZ	PA000	PAOZZ	30554	96-23534	SWITCH,THERMOSTATIC...................1
2	PAOZZ	PA000	PAOZZ	0BXW5	MSP675C	TRANSDUCER,MOTIONAL...................1
3	PAOZZ	PA000	PAOZZ	30327	204SAE	VALVE,ANGLE.............................1
4	PAOZZ	PA000	PAOZZ	30554	96-23640	ADAPTER,STRAIGHT....................1
5	PAOZZ	PA000	PAOZZ	93061	222P-4-2	ADAPTER................................1
6	PAOZZ	PA000	PAOZZ	06YD3	MSP300-100-P-3-N	TRANSMITTER,PRESSUR...................1
7	PAOZZ	PA000	PAOZZ	19617	HW310	NUT,PLAIN,ASSEMBLED..................4
8	PAOZZ	PA000	PAOZZ	30554	88-20544-4	CLAMP,LOOP..........................2
9	XBOZZ	XB	XBOZZ	30554	96-23555	BRACKET,MOUNTING...................1
10	PAOZZ	PA000	PAOZZ	30554	88-20260-2	SCREW,CAP,HEXAGON H..................1
11	PAOZZ	PA000	PAOZZ	81640	8906K4522	SWITCH,TOGGLE......................1
12	XBOZZ	XB	XBOZZ	30554	88-21776	PLATE,DEAD CRANK 1
13	XBOZZ	XB	XBOZZ	81349	M24243/6-A402H	RIVET,BLIND..........................2
14	PAOZZ	PA000	PAOZZ	48370	933-M10	SCREW,CAP,HEXAGON H..................3
15	PAOZZ	PA000	PAOZZ	30554	96-23613-2	WASHER,SPLIT........................3
16	PAOZZ	PA000	PAOZZ	15526	125AZY-10M	WASHER,FLAT..........................3
17	PAOZZ	PA000	PAOZZ	30554	88-20546-4	CLAMP,LOOP...........................1
18	PAOZZ	PA000	PAOZZ	30554	88-22790-1	NUT...................................2
19	PAOZZ	PA000	PAOZZ	97403	13230E6744-63	WASHER,FLAT.........................2
20	PAOZZ	PA000	PAOZZ	96906	MS27183-10	WASHER,FLAT..........................2
21	XBOZZ	XB	XBOZZ	30554	88-22161-2	CABLE ASSEMBLY......................1
22	PAOZZ	PA000	PAOZZ	30554	88-20260-30	SCREW,CAP,HEXAGON H.................2
23	XBOZZ	XB	XBOZZ	30554	88-22162	BRACKET,TIE DOWN....................1
24	PAOZZ	PA000	PAOZZ	96906	MS51846-1	NIPPLE,PIPE........................1
25	XBOZZ	XB	XBOZZ	30554	88-22481	PLATE,IDENTIFICATIO................1
26	PAOZZ	PA000	PAOZZ	70411	SP2529VT	VALVE,CHECK.........................1
27	PAOZZ	PA000	PAOZZ	06383	PLT2S	STRAP,TIEDOWN,ELECT.................V
28	PAFZZ	PA000	PAFZZ	81343	2-4-070202	ELBOW,PIPE..........................1
29	PAFZZ	PA000	PAFZZ	46717	LA-519-9	COUPLING,PIPE.......................1
30	PAFZZ	PA000	PAFZZ	0BXW5	ADC100-24	ACTUATOR,ELECTRO-ME.................1
31	PAOZZ	PA000	PAOZZ	15526	125AZY-5M	WASHER,FLAT.........................2
32	PAOZZ	PA000	PAOZZ	15526	127BZY-5M	WASHER,SPLIT........................2
33	PAOZZ	PA000	PAOZZ	12066	CTG-4635	TRANSFORMER,CURRENT.................1
34	PAOZZ	PA000	PAOZZ	96906	MS25036-110	TERMINAL,LUG........................1
35	PAOZZ	PA000	PAOZZ	019L2	21NTE82	NUT,SELF-LOCKING....................2
36	XBOZZ	XB	XBOZZ	30554	96-23671	BRACKET,MOUNTING....................1

END OF FIGURE

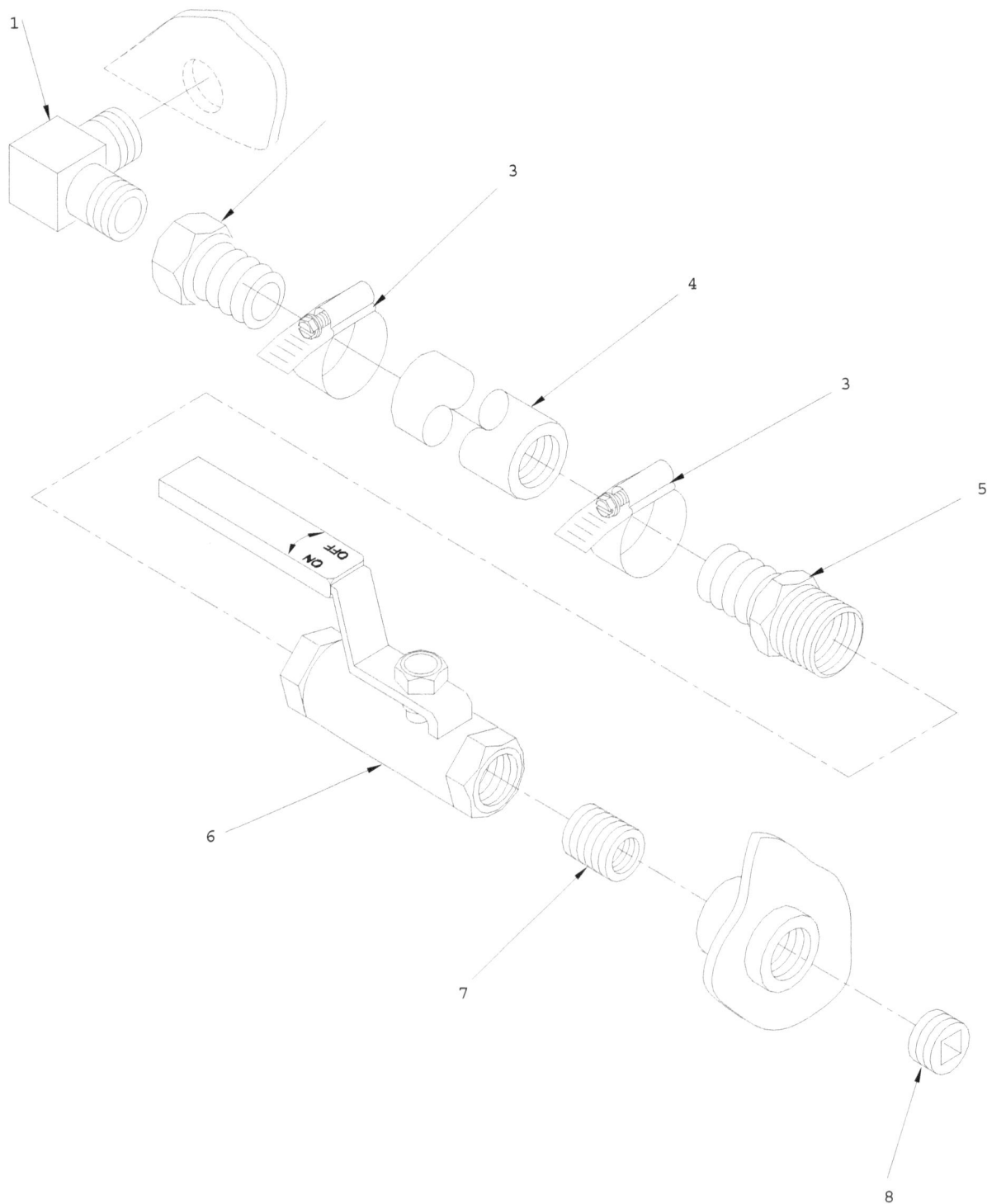

Figure 20. Lubrication System

(1)		(2)		(3)	(4)	(5)	(6)
					SMR CODE		
ITEM		AIR			PART	DESCRIPTION AND	
NO	ARMY	FORCE	USMC	CAGEC	NUMBER	USABLE ON CODE (UOC)	QTY

GROUP 09 LUBRICATION SYSTEM
FIGURE 20 LUBRICATION SYSTEM

1	PAOZZ	PA000	PAOZZ	30780	10M18C80MX	ELBOW,PIPE TO HOSE....................1
2	PAOZZ	PA000	PAOZZ	98441	30682-10-10B	ADAPTER,STRAIGHT,TU...................1
3	PAOZZ	PA000	PAOZZ	30554	88-20561-2	CLAMP,HOSE...........................2
4	PAOZZ	PA000	PAOZZ	30554	96-23514-1	HOSE,NONMETALLIC.....................1
5	PAOZZ	PA000	PAOZZ	93061	125HBL-10-12	ADAPTER,STRAIGHT,PI..................1
6	PAOZZ	PA000	PAOZZ	93061	V500P-12	VALVE,BALL...........................1
7	PAOZZ	PA000	PAOZZ	30554	72-5304	NIPPLE,PIPE..........................1
8	PAOZZ	PA000	PAOZZ	04826	12-130109G	PLUG,PIPE............................1

END OF FIGURE

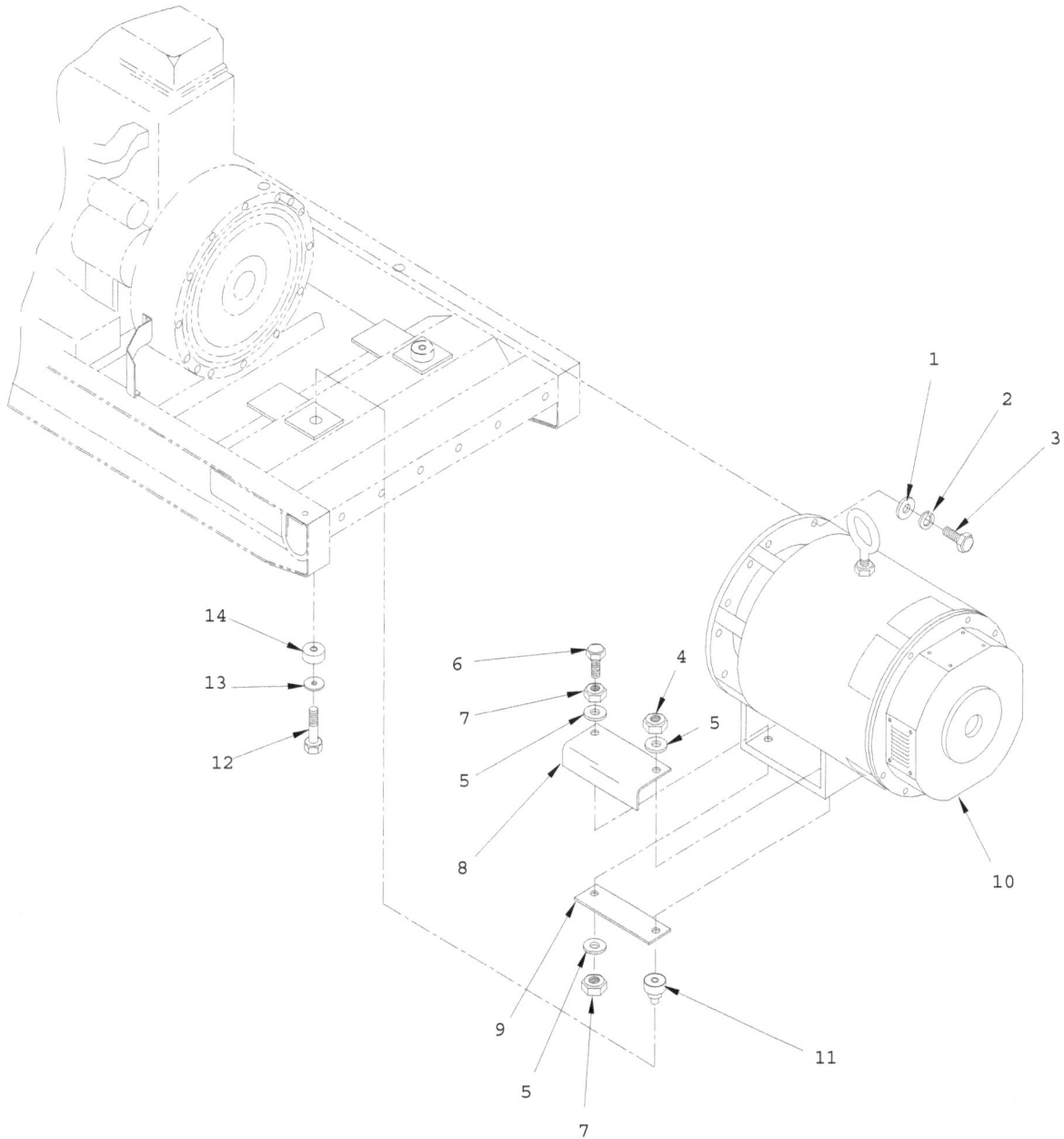

Figure 21. Generator Assembly (Sheet 1 of 2)

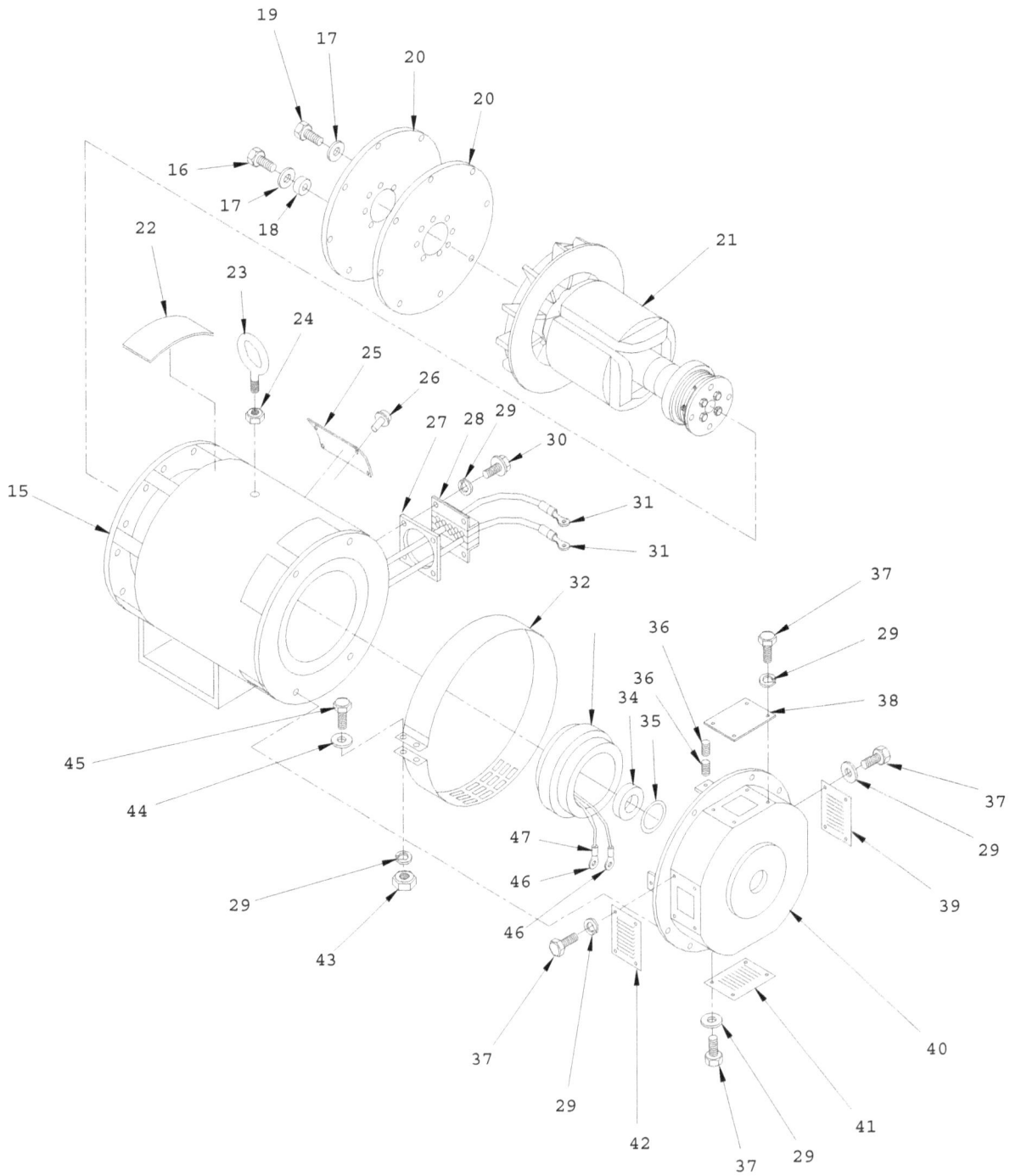

Figure 21. Generator Assembly (Sheet 2 of 2)

(1)	(2)			(3)	(4)	(5)	(6)
					SMR CODE		
ITEM	AIR				PART	DESCRIPTION AND	
NO	ARMY	FORCE	USMC	CAGEC	NUMBER	USABLE ON CODE (UOC)	QTY

GROUP 10 GENERATOR ASSEMBLY
FIGURE 21 GENERATOR ASSEMBLY

1	PAFZZ	PA000	PAFZZ	15526	125AZY-10M	WASHER,FLAT.........................20
2	PAFZZ	PA000	PAFZZ	15526	127BZY-10M	WASHER,SPLIT........................20
3	PAFZZ	PA000	PAFZZ	48370	933-M10	SCREW,CAP,HEXAGON H.................20
4	PAFZZ	PA000	PAFZZ	30554	88-20568-6	NUT,SELF-LOCKING.....................4
5	PAOZZ	PA000	PAOZZ	96906	MS51412-11	WASHER,FLAT.........................8
6	PAFZZ	PA000	PAFZZ	30554	88-22530	SCREW,CAP,HEXAGON H.................4
7	PAFZZ	PA000	PAFZZ	30554	88-22790-1	NUT.................................4
8	XBFZZ	XB	XBFZZ	30554	88-22527	ANGLE,MOUNT.........................2
9	XBFZZ	XB	XBFZZ	30554	88-22526	PLATE,MOUNT.........................2
10	PBFFF	PB000	PBFFF	38151	88-21007	GENERATOR,ALTERNATO.................1
						UOC: LTY
10	PBFFF	PB000	PBFFF	38151	88-21008	GENERATOR,ALTERNATO.................1
						UOC: LTX
11	PAFZZ	PA000	PAFZZ	81860	RB-X86	MOUNT,RESILIENT.....................2
12	PAFZZ	PA000	PAFZZ	80204	B1821BH063C475N	SCREW,CAP,HEXAGON H.................2
13	PAOZZ	PA000	PAOZZ	92830	B1875-127	WASHER,BEVEL........................2
14	PAOZZ	PA000	PAOZZ	30554	88-20247-3	WASHER,FLAT.........................2
15	XBFZZ	XB	XBFZZ	36156	778663-0B	FRAME,STATOR ASSEMB,400 HZ..........1
						UOC: LTX
15	XAFZZ	XB	XAFZZ	30554	88-21922	FRAME,STATOR ASSEMBLY,60 HZ.........1
						UOC: LTY
16	PAFZZ	PA000	PAFZZ	80204	B1821BH038F175N	SCREW,CAP,HEXAGON H.................2
17	PAFZZ	PA000	PAFZZ	96906	MS27183-13	WASHER,FLAT.........................8
18	XBFZZ	XB	XBFZZ	30554	88-22205	SPACER,COUPLER......................2
19	PAFZZ	PA000	PAFZZ	96906	MS90726-60	SCREW,CAP,HEXAGON H.................6
20	XBFZZ	XB	XBFZZ	36156	702807-01	DISC,COUPLING.......................2
21	PAFFF	PA000	PAFFF	36156	777066-0A	ROTOR ASSEMBLY,60 HZ................1
						(SEE FIGURE 23 FOR PARTS BREAKDOWN)
						UOC: LTY
21	PAFFF	PA000	PAFFF	36156	777067-0A	ROTOR ASSEMBLY,400 HZ...............1
						(SEE FIGURE 23 FOR PARTS BREAKDOWN)
						UOC: LTX
22	XBFZZ	XB	XBFZZ	36156	A-9696	LABEL,WARNING.......................1
23	PAOZZ	PA000	PAOZZ	96906	MS51937-7	BOLT,EYE............................1
24	PAFZZ	PA000	PAFZZ	96906	MS51967-20	NUT,PLAIN,HEXAGON...................1
25	XBFZZ	XB	XBFZZ	30554	88-20064-7	PLATE,IDENTIFICATIO,60 HZ...........1
						UOC: LTY
25	XBFZZ	XB	XBFZZ	30554	88-20064-8	PLATE,IDENTIFICATIO,400 HZ
						UOC: LTX
26	XBFZZ	XB	XBFZZ	80205	MS21318-21	SCREW,DRIVE.........................4
27	XBFZZ	XB	XBFZZ	36156	834822-01	GASKET,LEAD CLAMP...................1
28	XBFZZ	XB	XBFZZ	36156	725819-0A	LEAD CLAMP ASSEMBLY.................1

(1)	(2)			(3)	(4)	(5)	(6)
					SMR CODE		
ITEM	AIR				PART	DESCRIPTION AND	
NO	ARMY	FORCE	USMC	CAGEC	NUMBER	USABLE ON CODE (UOC)	QTY
29	PAFZZ	PA000	PAFZZ	96906	MS35338-44	WASHER,LOCK...........................24	
30	PAFZZ	PA000	PAFZZ	96906	MS90725-6	SCREW,CAP,HEXAGON H...................4	
31	PAFZZ	PA000	PAFZZ	96906	MS20659-129	TERMINAL,LUG..........................1	
32	XBFZZ	XB	XBFZZ	36156	720510-0B	COVER,BAND............................2	
						UOC: LTY	
32	XBFZZ	XB	XBFZZ	36156	720982-0A	COVER,BAND............................2	
						UOC: LTX	
33	PBFZZ	PB000	PBFZZ	36156	789293-0A	STATOR,GENERATOR......................1	
34	PAFZZ	PA000	PAFZZ	92862	A7812-60	BEARING,BALL,ANNULA...................1	
35	PAFZZ	PA000	PAFZZ	36156	865873-01	O-RING................................1	
36	PAFZZ	PA000	PAFZZ	36156	801048-04	SETSCREW..............................8	
37	PAFZZ	PA000	PAFZZ	96906	MS90725-3	SCREW,CAP,HEXAGON H..................16	
38	XBFZZ	XB	XBFZZ	36156	718514-01	COVER,TOP.............................1	
39	XBFZZ	XB	XBFZZ	36156	718515-01	COVER,LOUVERED,RS.....................1	
40	XBFZZ	XB	XBFZZ	36156	703512-01	ENDBELL...............................1	
41	XBFZZ	XB	XBFZZ	36156	718517-01	COVER,LOUVERED,BO.....................1	
42	XBFZZ	XB	XBFZZ	36156	718516-01	COVER,LOUVERED,LS.....................1	
43	PAFZZ	PA000	PAFZZ	1FP59	012316-7	NUT,PLAIN,HEXAGON.....................4	
44	PAFZZ	PA000	PAFZZ	96906	MS27183-49	WASHER,FLAT...........................5	
45	PAFZZ	PA000	PAFZZ	96906	MS35206-287	SCREW,MACHINE.........................4	
46	PAFZZ	PA000	PAFZZ	96906	MS20659-165	TERMINAL,LUG..........................2	
47	M00ZZ	M000	M0000	30554	88-21007-36	INSULATION,SLEEVING...................2	
						MAKE FROM P/N M3190/3-17-0 (81349),	

2 INCH

END OF FIGURE

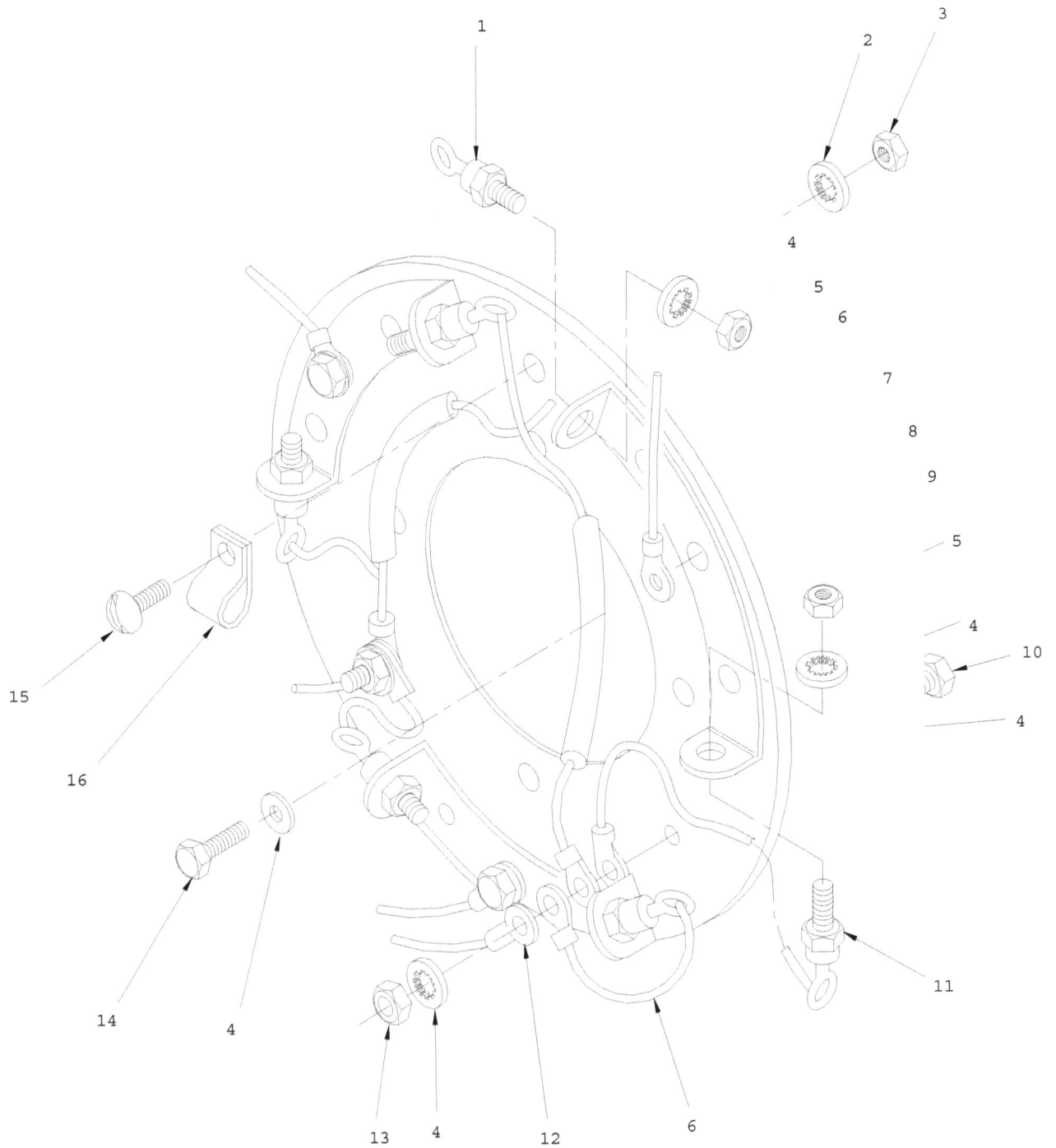

Figure 22. Rotating Rectifier

(1)				(3)	(4)	(5)	(6)
		(2)			SMR CODE		
ITEM		AIR			PART	DESCRIPTION AND	
NO	ARMY	FORCE	USMC	CAGEC	NUMBER	USABLE ON CODE (UOC)	QTY

GROUP 1002 ROTATING RECTIFIER
FIGURE 22 ROTATING RECTIFIER

1	PAFZZ	PA000	PAFZZ	81349	JANTX1N1190R	SEMICONDUCTOR DEVIC...................3
2	PAFZZ	PA000	PAFZZ	96906	MS35333-38	WASHER,LOCK...........................1
3	PAFZZ	PA000	PAFZZ	96906	MS35649-282	NUT,PLAIN,HEXAGON.....................1
4	PAFZZ	PA000	PAFZZ	96906	MS35333-40	WASHER,LOCK..........................21
5	PAFZZ	PA000	PAFZZ	88044	AN315-4R	NUT,PLAIN,HEXAGON.....................6
6	MFFZZ	M000	MFFZZ	30554	88-21649-19	WIRE,ELECTRICAL.......................6

MAKE FROM P/N 88-20444-2 (30554)
AS REQD

7	XBFZZ	XB	XBFZZ	36156	B-718930-01	PLATE,RECTIFIER.......................2
8	XBFZZ	XB	XBFZZ	36156	B-718817-01	PLATE,RECTIFIER.......................1
9	MFFZZ	M000	MFFZZ	30554	88-21649-87	INSULATION,SLEEVING...................2

MAKE FROM P/N M3190/3-17-0 (81349),
7 INCH

10	PAFZZ	PA000	PAFZZ	96906	MS35206-203	SCREW,MACHINE.........................2
11	PAFZZ	PA000	PAFZZ	81349	JANTX1N1190	SEMICONDUCTOR DEVIC...................3
12	PAFZZ	PA000	PAFZZ	96906	MS25036-154	TERMINAL,LUG..........................6
13	PAFZZ	PA000	PAFZZ	1FP59	012316-7	NUT,PLAIN,HEXAGON.....................4
14	PAFZZ	PA000	PAFZZ	96906	MS90725-3	SCREW,CAP,HEXAGON H...................9
15	PAFZZ	PA000	PAFZZ	96906	MS35206-248	SCREW,MACHINE.........................1
16	PAFZZ	PA000	PAFZZ	96906	MS25281-4	CLAMP,LOOP............................1

END OF FIGURE

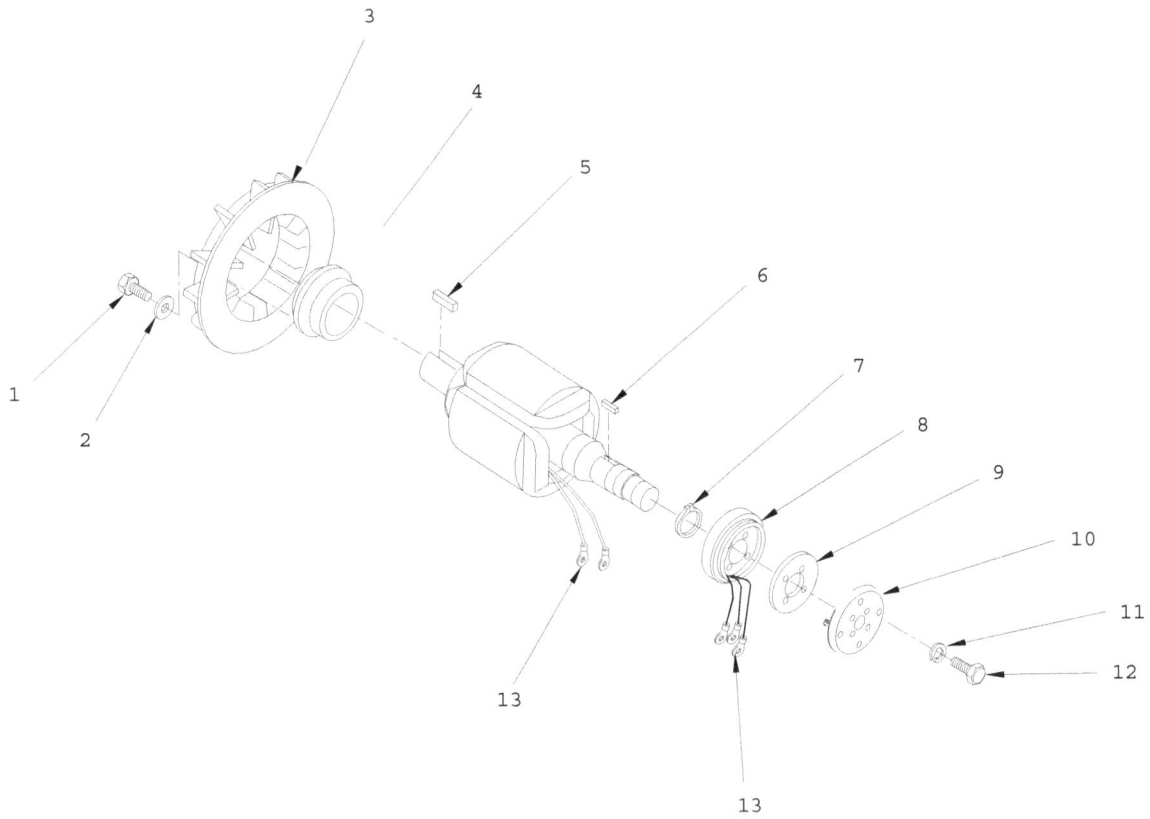

Figure 23.   Generator Rotor Assembly

(1)		(2)		(3)	(4) SMR CODE	(5)	(6)
ITEM NO	ARMY	AIR FORCE	USMC	CAGEC	PART NUMBER	DESCRIPTION AND USABLE ON CODE (UOC)	QTY

GROUP 1005 GENERATOR ROTOR ASSEMBLY
FIGURE 23 GENERATOR ROTOR ASSEMBLY

1	PAFZZ	PA000	PAFZZ	80204	B1821BH038C125N	SCREW,CAP,HEXAGON H...................4
2	PAFZZ	PA000	PAFZZ	96906	MS35338-46	WASHER,LOCK..........................4
3	XBFZZ	XB	XBFZZ	36156	716511-02	FAN..................................1
4	XBFZZ	XB	XBFZZ	36156	707509-01	HUB,DRIVE............................1
5	PAFZZ	PA000	PAFZZ	96906	MS20066-358	KEY,MACHINE..........................1
6	PAFZZ	PA000	PAFZZ	96906	MS20066-356	KEY,MACHINE..........................1
7	PAFZZ	PA000	PAFZZ	96906	MS16624-3275	RING,RETAINING.......................1
						UOC: LTY
7	PAFZZ	PA000	PAFZZ	36156	832805-01	RING,RETAINING.......................1
						UOC: LTX
8	PAFZZ	PA000	PAFZZ	36156	791150-0A	EXCITER,ARMATURE.....................1
9	PAFZZ	PA000	PAFZZ	36156	707807-02	RECTIFIER,METALLIC...................1
10	PBFFF	PB000	PBFFF	36156	777056-0A	RECTIFIER ASSEMBLY...................1
11	XAFZZ	PA000	XAFZZ	96906	MS35333-40	WASHER,LOCK..........................4
12	XAFZZ	PA000	XAFZZ	96906	MS90725-2	SCREW,CAP,HEXAGON....................4
						UOC: LTY
12	PAFZZ	PA000	PAFZZ	96906	MS90725-5	SCREW,CAP,HEXAGON....................4
						UOC: LTX
13	PAFZZ	PA000	PAFZZ	96906	MS25036-157	TERMINAL,LUG.........................5

END OF FIGURE

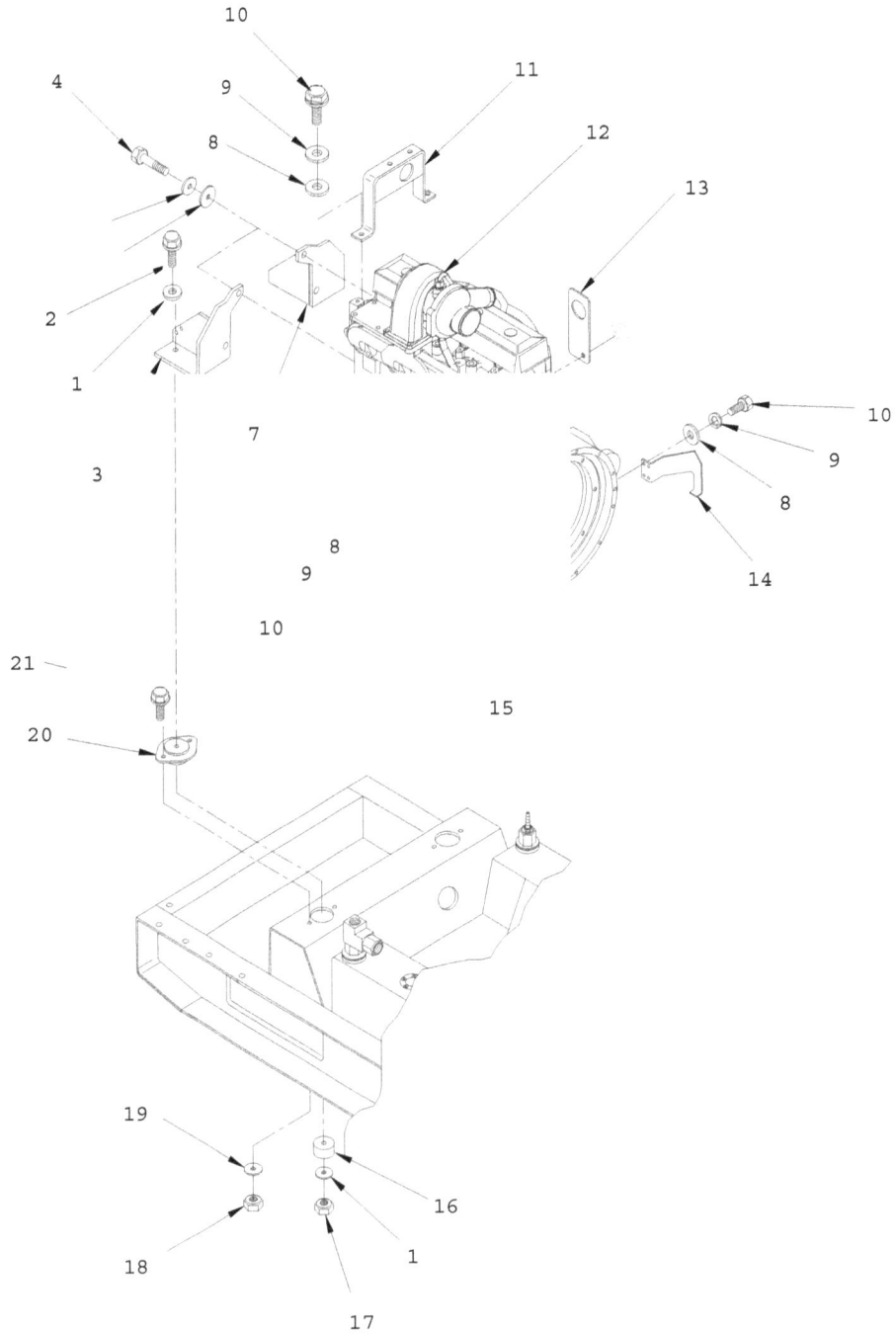

Figure 24. Engine Assembly

(1)		(2)		(3)	(4) SMR CODE PART	(5)	(6)
ITEM NO	ARMY	AIR FORCE	USMC	CAGEC	NUMBER	DESCRIPTION AND USABLE ON CODE (UOC)	QTY

GROUP 11 ENGINE ASSEMBLY
FIGURE 24 ENGINE ASSEMBLY

Item	Army	Air Force	USMC	CAGEC	Part Number	Description	Qty
1	PAFZZ	PA000	PAFZZ	96906	MS51412-9	WASHER,FLAT	2
2	PAFZZ	PA000	PAFZZ	80204	B1821BH050C300N	SCREW,CAP,HEXAGON H	2
3	XB0ZZ	XB	XB0ZZ	30554	96-23528	BRACKET,MOUNTING	1
4	PAFZZ	PA000	PAFZZ	15526	933-ZY-M16-50	BOLT,MACHINE	4
5	PAFZZ	PA000	PAFZZ	15526	127BZY-16M	WASHER,SPLIT	4
6	PAFZZ	PA000	PAFZZ	15526	125AZY-16M	WASHER,FLAT	4
7	XBFZZ	XB	XBFZZ	30554	88-22020	SUPPORT ASSEMBLY,EN	1
8	PAFZZ	PA000	PAFZZ	15526	125AZY-16M	WASHER,FLAT	4
9	PAFZZ	PA000	PAFZZ	15526	127BZY-12M	WASHER,SPLIT	4
10	PAFZZ	PA000	PAFZZ	48370	933-M12	SCREW,CAP,HEXAGON H	4
11	XBFZZ	XB	XBFZZ	30554	96-23522	SUPPORT,STRUCTURAL	1
12	PB0HH	PB000	PB0HH	66836	T04045TF151	ENGINE,DIESEL (SEE TM 9-2815-259-24P FOR PARTS BREAKDOWN)	1
13	XBFZZ	XB	XBFZZ	30554	88-22117	STRAP,LIFTING	1
14	XBFZZ	XB	XBFZZ	30554	88-22130	SUPPORT,ENGINE	1
15	XBFZZ	XB	XBFZZ	30554	88-21882	SUPPORT,ENGINE	1
16	PAFZZ	PA000	PAFZZ	30554	88-20247-2	WASHER,FLAT	2
17	PAFZZ	PA000	PAFZZ	30554	88-20568-5	NUT,SELF-LOCKING	2
18	PAFZZ	PA000	PAFZZ	30554	88-20568-4	NUT,SELF-LOCKING	4
19	PAFZZ	PA000	PAFZZ	96906	MS21306-1G	WASHER,FLAT	1
20	PAFZZ	PA000	PAFZZ	30554	96-23607	MOUNT,RESILIENT	2
21	PAFZZ	PA000	PAFZZ	80204	B1821BH038C100N	SCREW,CAP,HEXAGON H	1

END OF FIGURE

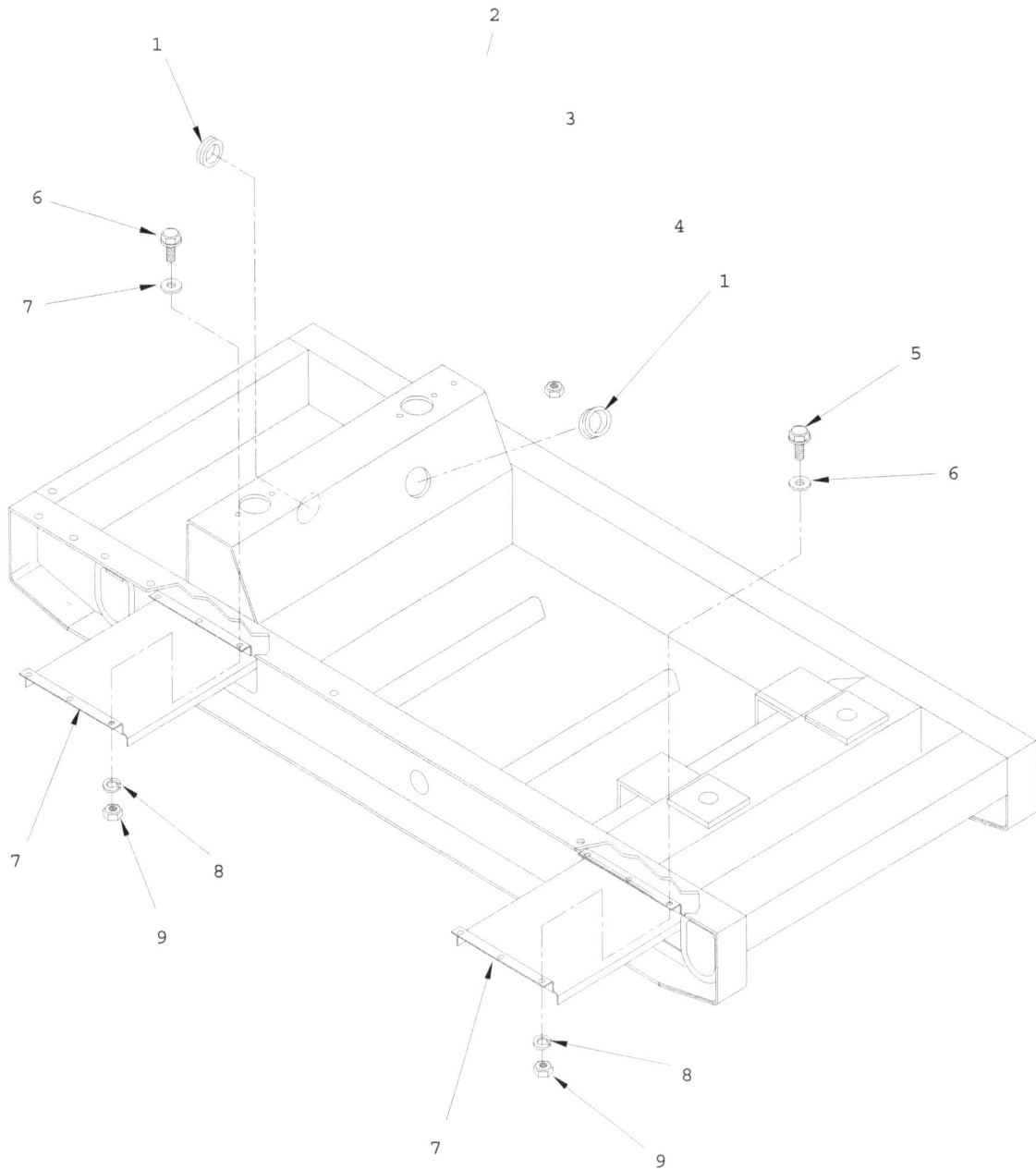

Figure 25. Skid Base

(1) ITEM NO	ARMY	(2) AIR FORCE	USMC	(3) CAGEC	(4) SMR CODE PART NUMBER	(5) DESCRIPTION AND USABLE ON CODE (UOC)	(6) QTY

GROUP 12 SKID BASE
FIGURE 25 SKID BASE

(1)		(2)		(3)	(4)	(5)	(6)
1	PAFZZ	PAOOO	PAFZZ	70485	2774	GROMMET,NONMETALLIC...................1
2	PAFZZ	PAOOO	PAFZZ	30554	88-20260-23	SCREW,CAP,HEXAGON H..................1
3	PAOZZ	PAOOO	PAOZZ	22175	55MC1-11-SS-R	CLAMP,LOOP...........................1
4	PAFZZ	PAOOO	PAFZZ	19617	HW310	NUT,PLAIN,ASSEMBLED..................1
5	PAFZZ	PAOOO	PAFZZ	73801	MS35489-88	GROMMET,NONMETALLIC..................1
6	PAFZZ	PAOOO	PAFZZ	80204	B1821BH038C100N	SCREW,CAP,HEXAGON H..................6
7	PAFZZ	PAOOO	PAFZZ	96906	MS21306-1G	WASHER,FLAT..........................6
8	PAFZZ	PAOOO	PAFZZ	97403	13230E6744-65	WASHER,LOCK..........................1
9	PAFZZ	PAOOO	PAFZZ	30554	88-22790-3	NUT,PLAIN,HEXAGON....................6
10	XBFZZ	XB	XBFZZ	30554	88-21731	GUIDE,FORKLIFT.......................2

END OF FIGURE

(1)	(2)			(3)	(4)	(5)	(6)
					SMR CODE		
ITEM	AIR				PART	DESCRIPTION AND	
NO	ARMY	FORCE	USMC	CAGEC	NUMBER	USABLE ON CODE (UOC)	QTY

GROUP 99 BULK ITEMS

1	PAOZZ	PA000	PAOZZ	99739	55-1986-19	HOSE,NONMETALLIC......................V
2	PAOZZ	PA000	PAOZZ	5G996	1/0SGT	CABLE,POWER,ELECTRI...................V
3	PAOZZ	PA000	PAOZZ	98441	208-4	HOSE,NONMETALLIC......................V
4	PAOZZ	PA000	PAOZZ	81349	M5086/2-16-9	WIRE,ELECTRICAL.......................V
5	PAOZZ	PA000	PAOZZ	98441	208-5	HOSE,NONMETALLIC......................V
6	PAFZZ	PA000	PAFZZ	1X968	ZX-4134	SEAL,NONMETALLIC,SP...................V
7	PAOZZ	PA000	PAOZZ	81349	MIL-H-13444,TYPE	HOSE,NONMETALLIC......................V
8	PAOZZ	PA000	PAOZZ	28818	FF40JM02	SOUND CONTROLLING B...................V
						POLYETHER URETHANE FOAM, GREY, 1.0
						IN THK
9	PAOZZ	PA000	PAOZZ	30554	88-21908	HOSE,NONMETALLIC......................V
						CLASS D-2, 1.75 IN ID
10	PAOZZ	PA000	PAOZZ	98441	160-12	HOSE,NONMETALLIC......................V
						MIL-H-6000-12, 0.75 NOM ID
11	PAOZZ	PA000	PAOZZ	4T483	00.625TF12-50	HOSE,METALLIC.........................V
12	PAOZZ	PA000	PAOZZ	52152	1430	TAPE,SHIELDING,ELEC...................V
13	PAOZZ	PA000	PAOZZ	03938	APT11812	INSULATION,SLEEVING...................V

END OF FIGURE

NATIONAL STOCK NUMBER INDEX

STOCK NUMBER	FIG.	ITEM	STOCK NUMBER	FIG.	ITEM
5306-00-017-6143	21	23	5310-00-167-1344	22	5
4720-00-021-3320	1	22	5325-00-174-9038	10	19
5305-00-036-6902	5	70	5310-00-189-8467	17	4
	14	13	5305-00-191-6226	7	35
5305-00-036-6972	4	125		14	20
4730-00-041-2526	11	54	4730-00-196-1991	7	28
5310-00-044-6477	10	6	4730-00-200-0531	11	56
	11	58	5305-00-211-9344	4	95
5310-00-052-3632	5	37	5305-00-224-1092	5	36
	11	25		14	47
	14	16	5306-00-226-4827	10	5
5310-00-063-7360	4	96		11	62
	5	21	5940-00-230-0515	22	12
5305-00-068-0500	21	37	4730-00-230-1996	19	24
	22	14	4730-00-234-7637	20	2
5305-00-068-0501	23	12	5940-00-237-2703	17	17
5305-00-068-0502	21	30	5305-00-269-2803	21	19
5305-00-068-0508	10	59	5305-00-269-3241	21	16
5305-00-068-0509	17	14	4030-00-272-9002	10	13
5305-00-068-0510	17	31	4730-00-288-9928	11	69
	24	21	4820-00-289-3378	19	3
	25	6	5330-00-290-9891	5	47
5305-00-068-0511	10	33	5340-00-297-0312	4	68
	23	1		17	30
5305-00-071-2075	24	2	5935-00-315-9563	15	9
5305-00-071-2514	14	38	5325-00-319-0547	25	1
5325-00-074-3301	4	38	5975-00-371-9428	4	93
5310-00-087-7493	21	17	4730-00-415-3172	19	29
6645-00-089-8842	5	18	5945-00-458-3351	5	76
5975-00-111-3208	5	80	5935-00-482-7721	12	4
	6	6	5306-00-484-5730	4	40
	15	4		5	94
5940-00-113-0954	21	46		11	21
5940-00-113-8191	14	51	5310-00-498-7234	2	16
	16	4		4	20
	16	8		5	5
	16	12		5	79
5940-00-113-9831	14	53		10	11
	16	16		14	6
	16	20		14	12
	16	24		17	11
5940-00-114-1314	21	31		19	7
2590-00-141-9758	11	12		25	4
5940-00-143-4777	23	13	5305-00-543-4372	7	20
5940-00-143-4780	6	7	5940-00-549-6581	2	2
	15	14	5940-00-549-6583	2	4
5940-00-143-4793	15	17	5310-00-550-1130	22	4
	19	34	5310-00-559-0070	22	2
5961-00-154-7046	22	11	6145-00-578-6605	BULK	4
3110-00-155-6298	21	34	5310-00-582-5965	21	29

NATIONAL STOCK NUMBER INDEX

STOCK NUMBER	FIG.	ITEM	STOCK NUMBER	FIG.	ITEM
6210-00-583-9349	5	7	5305-01-056-1501	4	56
4730-00-595-3108	7	26		7	5
4730-00-613-6468	11	40	5905-01-063-9644	17	26
5930-00-615-6731	5	17	4810-01-067-8276	13	1
5310-00-637-9541	23	2	5961-01-067-9493	22	1
5330-00-641-4336	5	50	5980-01-074-1633	5	10
5930-00-655-4247	5	14	5340-01-078-9038	5	99
5305-00-721-5492	2	14	5940-01-082-3321	5	63
5305-00-724-7265	21	12		6	14
5305-00-725-2317	10	4		15	20
5310-00-761-6882	21	43		18	3
	22	13		18	7
5310-00-763-8920	21	24		18	11
4730-00-765-9103	11	39		18	15
4820-00-785-8153	10	24	5999-01-092-2655	12	3
	11	33	5935-01-097-9974	2	17
4720-00-804-9249	11	68	5940-01-112-9746	7	32
5310-00-809-4058	1	17		7	38
	4	19		13	2
	7	3		15	21
	10	57	5940-01-126-3973	7	33
	11	63		7	39
	14	39		13	3
	17	16	5975-01-128-0390	6	4
	19	20		12	2
4730-00-812-1333	11	23		15	3
4730-00-821-7917	11	4		19	27
5310-00-822-8525	4	61	4730-01-134-9827	11	18
	5	89	5940-01-139-0853	5	62
5310-00-836-3520	5	34		6	15
	14	46	5306-01-156-7663	2	18
5315-00-847-3531	23	6		4	25
5935-00-852-9611	15	19		5	4
5975-00-879-7234	5	46		5	81
5305-00-889-3118	22	10		10	14
5920-00-892-9311	5	30		14	3
5310-00-903-8595	4	5		14	10
	14	17		17	8
5325-00-907-1183	25	5	5325-01-161-2654	5	90
4720-00-913-5910	BULK	7	5305-01-187-5878	14	33
5310-00-934-9757	22	3	4820-01-192-7676	8	5
6210-00-935-6919	5	9	5315-01-222-9228	23	5
5975-00-944-1499	6	5	5340-01-223-4373	7	8
5940-00-954-3558	5	29	5310-01-234-9415	4	67
5305-00-984-6196	22	15	5310-01-234-9416	4	3
5305-00-988-1728	21	45		5	3
5340-00-988-3210	22	16		5	51
5935-01-005-9478	5	43		10	36
4820-01-008-2922	20	6		11	29
5999-01-039-8438	15	10		14	32

NATIONAL STOCK NUMBER INDEX

STOCK NUMBER	FIG.	ITEM	STOCK NUMBER	FIG.	ITEM
5310-01-242-2679	21	5	5340-01-368-5596	7	6
5305-01-247-6829	5	41	5306-01-368-8041	4	50
5940-01-259-2190	6	9	6620-01-369-0432	7	21
	15	18	5305-01-369-2166	4	17
5310-01-266-4641	24	1		19	22
5310-01-267-1685	11	16	5940-01-369-2267	6	8
	17	2		15	16
5970-01-280-0362	5	98	5940-01-369-2269	6	13
4730-01-296-9318	11	19	5940-01-369-2271	16	2
2920-01-298-6321	23	10		16	6
4730-01-301-4257	7	30		16	10
5325-01-301-7903	4	60		16	14
	5	84		16	18
5310-01-333-1883	21	44		16	22
5905-01-336-7533	5	32	5940-01-369-2872	3	2
4730-01-340-7594	20	5		3	7
5310-01-365-3190	24	16		3	12
5310-01-365-4381	5	88		3	17
5310-01-365-4386	21	13		3	21
5310-01-365-5788	17	1	5940-01-369-6948	18	4
5305-01-365-6313	4	4		18	8
	5	1		18	12
	11	15		18	16
	25	2	5330-01-369-7318	21	35
5305-01-365-6314	17	13	5330-01-369-8947	BULK	6
2940-01-365-6535	11	48	4720-01-369-9366	7	11
5310-01-365-8139	21	14	2930-01-370-2868	10	64
5930-01-365-9614	19	11	4730-01-370-5426	10	34
5945-01-365-9953	14	19		11	35
5999-01-366-2621	5	71	5340-01-370-8349	5	69
5330-01-366-2836	11	31	6110-01-372-2597	1	16
5342-01-366-3361	21	11		14	44
5305-01-366-3501	21	6	5930-01-372-5387	5	12
5320-01-366-4394	4	8	4720-01-373-0526	BULK	9
5310-01-366-4412	17	34	5307-01-374-4451	4	58
2930-01-366-6107	10	54	4720-01-375-1392	11	37
5306-01-366-7075	4	66	4720-01-375-1930	7	15
2910-01-366-7293	12	1	5120-01-375-4373	17	9
5310-01-366-8134	5	26	5930-01-376-2001	5	11
2910-01-366-8983	11	30		5	13
5342-01-367-1516	24	20	2910-01-377-3984	11	57
4820-01-367-1836	19	26	9905-01-377-5094	1	6
5935-01-367-4422	5	68	2940-01-378-1130	8	2
5935-01-367-7814	5	45	4730-01-378-5224	10	26
6110-01-367-8921	14	28	6140-01-378-8232	2	7
5340-01-367-8956	4	7	5930-01-379-0644	5	15
5905-01-368-2539	5	35	5950-01-382-3371	14	37
5930-01-368-2892	5	19	5999-01-382-8223	2	3
5950-01-368-2915	14	34	5940-01-384-0384	14	41
5930-01-368-5161	5	20	5910-01-384-1745	14	49

NATIONAL STOCK NUMBER INDEX

STOCK NUMBER	FIG.	ITEM	STOCK NUMBER	FIG.	ITEM
5340-01-384-6650	4	124	5310-01-466-6687	10	35
6115-01-385-3104	21	33		25	9
4720-01-386-1856	11	7	5310-01-466-7247	17	18
5640-01-386-9618	BULK	8	5925-01-466-8153	5	23
4730-01-387-4000	10	27	5305-01-466-9756	4	36
6115-01-387-4676	21	10	5340-01-467-0760	4	43
2920-01-388-2776	4	65	4140-01-467-1386	10	51
2990-01-389-3003	4	6	5305-01-467-1561	5	22
4720-01-392-0319	11	50	2920-01-467-2624	21	21
5915-01-394-0942	17	6	2920-01-467-2626	21	21
6115-01-395-6536	21	10	7025-01-467-4709	5	40
5310-01-396-5836	10	50	5325-01-469-9637	23	7
	19	35	5310-01-469-9836	4	2
5310-01-396-5840	2	11		5	2
	24	19		5	52
	25	7		7	34
5915-01-396-9253	17	22		11	28
5930-01-398-6692	5	27	5310-01-469-9864	4	18
5310-01-406-1672	4	106		5	95
	5	55		7	4
5330-01-406-5707	14	4		10	58
6150-01-406-9533	1	21		11	22
4730-01-407-0649	11	3		14	42
5310-01-407-7191	4	76		17	15
4730-01-414-4281	10	28		19	19
5940-01-425-2020	14	36	5310-01-469-9865	10	7
5365-01-431-4540	4	105		11	59
5961-01-431-4558	5	28	5310-01-469-9866	2	15
5365-01-431-4603	4	110		7	17
5640-01-434-1250	BULK	13		10	3
5930-01-453-2655	5	16		17	29
5920-01-459-3847	5	31		25	8
2815-01-462-2289	24	12	5310-01-470-0446	19	32
5305-01-464-6667	4	42	5310-01-470-0448	10	53
	5	24	5310-01-470-0450	24	9
5340-01-465-7765	11	14	5310-01-470-0455	24	5
5330-01-465-9273	11	13	5945-01-470-0491	11	65
6110-01-466-4723	5	6	5905-01-470-0564	5	67
5998-01-466-4724	5	58	5945-01-470-0691	11	66
6110-01-466-4725	5	6	4710-01-470-0876	11	32
5998-01-466-4726	5	54	9390-01-470-1205	10	61
5310-01-466-4926	17	19	5310-01-470-1286	10	8
5998-01-466-6097	5	58		11	60
5310-01-466-6312	4	41	5310-01-470-1370	19	31
	7	7	5310-01-470-1373	10	44
	10	66		19	16
	11	49		21	1
	19	18	5905-01-470-1376	5	74
	21	7	5310-01-470-1381	24	6
5310-01-466-6687	7	16		24	8

NATIONAL STOCK NUMBER INDEX

STOCK NUMBER	FIG.	ITEM	STOCK NUMBER	FIG.	ITEM
5305-01-470-1421	5	91	4720-01-470-3016	10	21
5305-01-470-1425	5	39	5310-01-470-3024	21	4
4730-01-470-1567	7	22	5940-01-470-3031	14	14
4730-01-470-1578	7	2	5306-01-470-3057	10	52
	11	51	5306-01-470-3069	24	10
4730-01-470-1595	10	9	5340-01-470-3070	5	59
4730-01-470-1626	10	17	5905-01-470-3203	5	33
	11	2	4720-01-470-3561	10	23
5970-01-470-1631	3	3	4730-01-470-3566	10	22
	3	8	4730-01-470-3579	19	4
	3	15	6620-01-470-3854	19	6
	3	23	6695-01-470-3873	19	2
5970-01-470-1637	3	5	5310-01-470-3877	5	42
	3	10	4720-01-470-3917	BULK	10
	3	14	4720-01-470-3929	BULK	3
	3	19	4730-01-470-3971	20	1
	3	24	5330-01-470-4206	10	48
4730-01-470-1701	11	26	6110-01-470-4245	5	38
2910-01-470-1941	19	30	6110-01-470-4253	5	61
5306-01-470-1976	24	4	6110-01-470-4263	5	61
5310-01-470-1981	1	18	6110-01-470-4266	5	60
	14	45	5995-01-470-4269	5	49
4730-01-470-1982	20	7	6240-01-470-4272	5	8
5310-01-470-1998	2	9	5305-01-470-4533	10	46
4720-01-470-2086	10	10	5961-01-470-4673	14	24
5306-01-470-2097	10	42	4720-01-470-4779	BULK	11
	19	14	5305-01-470-5134	5	93
	21	3	4730-01-470-5158	7	27
4710-01-470-2119	11	41	5342-01-470-5177	10	16
5310-01-470-2355	10	43	5340-01-470-5184	5	53
	19	15	4720-01-470-5994	BULK	1
	21	2	6150-01-470-6001	15	11
1560-01-470-2391	7	1	6150-01-470-6017	5	48
4730-01-470-2409	7	14	4730-01-470-6039	11	47
	20	3	4730-01-470-6041	11	5
5305-01-470-2429	4	112	6145-01-470-6195	BULK	2
5305-01-470-2440	4	107	4730-01-470-6199	11	53
5305-01-470-2448	11	27	5935-01-470-6221	5	44
6130-01-470-2451	5	101		14	48
5940-01-470-2470	5	73	5340-01-470-6228	19	8
5940-01-470-2768	14	15	4720-01-470-6230	BULK	5
5310-01-470-2776	24	17	5340-01-470-6235	14	11
5940-01-470-2875	5	66	5340-01-470-6324	5	78
5995-01-470-2881	5	56		19	17
5995-01-470-2882	5	57	5305-01-470-6335	4	91
5305-01-470-2911	5	77	4730-01-470-6338	19	28
5340-01-470-2929	5	102	6145-01-470-6341	15	7
5950-01-470-2938	14	35	5340-01-470-6357	7	10
5950-01-470-2963	19	33	2940-01-470-6444	9	1
5340-01-470-2973	5	97	5340-01-470-6499	25	3

NATIONAL STOCK NUMBER INDEX

STOCK NUMBER	FIG.	ITEM	STOCK NUMBER	FIG.	ITEM
5310-01-470-6529	24	18			
4140-01-470-6555	10	60			
4140-01-470-6558	10	40			
4020-01-470-6597	17	7			
5895-01-470-6702	5	64			
5895-01-470-6709	5	64			
2990-01-470-7119	7	9			
2910-01-470-7423	11	8			
5935-01-470-7643	15	8			
5935-01-470-7699	5	105			
5930-01-470-8799	19	1			
2930-01-471-5127	10	38			
6150-01-472-8402	5	75			
2910-01-474-8437	11	67			
4930-01-475-0388	11	52			
5310-01-475-2155	11	24			
4010-01-475-2324	10	15			

PART NUMBER INDEX

CAGEC	PART NUMBER	STOCK NUMBER	FIG.	ITEM
58536	A-A-55804-I-A	5975-00-371-9428	4	93
36156	A-9696		21	22
58536	AA-52425-2	5940-00-549-6583	2	4
98410	AA-8714-08	5940-01-369-2269	6	13
0BXW5	ADC100-24	2910-01-470-1941	19	30
88044	AN315-4R	5310-00-167-1344	22	5
03938	APT11812	5640-01-434-1250	BULK	13
81343	ASTM A 466	4010-01-475-2324	10	15
0BXW5	AVR100	6110-01-470-4253	5	61
60886	AYW401-R	5930-01-379-0644	5	15
98441	A3411-54	4720-01-375-1392	11	37
61000	A3912-832-0	5340-01-470-2929	5	102
58536	A52425-1	5940-00-549-6581	2	2
61000	A7416-1032-0	5340-01-470-5184	5	53
61000	A7436-1032-0	5340-01-470-3070	5	59
92862	A7812-60	3110-00-155-6298	21	34
36156	B-718817-01		22	8
36156	B-718930-01		22	7
98410	B-801-12HDX	5940-01-369-6948	18	4
			18	8
			18	12
			18	16
98410	BB-818-38	5940-00-143-4793	15	17
98410	BB-8194-08	5940-01-259-2190	6	9
			15	18
98410	BB-837-10	5940-00-143-4780	6	7
			15	14
98410	BB-837-11		15	6
98410	BB-837-12		15	15
98410	BB-8707-06	5940-01-369-2267	6	8
			15	16
63857	BS4-1801PC15	5310-00-189-8467	17	4
80204	B1821BA025C225N	5305-01-470-6335	4	91
80204	B1821BH025C075N	5305-00-068-0508	10	59
80204	B1821BH025C125N	5305-00-068-0509	17	14
80204	B1821BH025C275N	5305-00-071-2514	14	38
80204	B1821BH031C100N	5306-00-226-4827	10	5
			11	62
80204	B1821BH038C063N	5305-00-721-5492	2	14
80204	B1821BH038C075N	5305-00-543-4372	7	20
80204	B1821BH038C100N	5305-00-068-0510	17	31
			24	21
			25	6
80204	B1821BH038C125N	5305-00-068-0511	10	33
			23	1
80204	B1821BH038C150N	5305-00-725-2317	10	4
80204	B1821BH038F175N	5305-00-269-3241	21	16
80204	B1821BH050C300N	5305-00-071-2075	24	2
80204	B1821BH063C475N	5305-00-724-7265	21	12
92830	B1875-127	5310-01-365-4386	21	13
1E045	CAT370		4	121

PART NUMBER INDEX

CAGEC	PART NUMBER	STOCK NUMBER	FIG.	ITEM
1E045	CAT380		4	122
61529	CB1AF-M-24V	5945-01-470-0691	11	66
14407	CC1-00380-110L	5950-01-470-2938	14	35
74545	CR15	5935-01-367-7814	5	45
12066	CTG-4635	5950-01-470-2963	19	33
10983	C10-C57410E	5930-01-368-5161	5	20
79470	C3059X2		11	10
23040	C391674	4730-00-288-9928	11	69
78553	C7931-1032-3B	5310-00-903-8595	4	5
			14	17
78553	C7941-1032-3B	5310-01-366-8134	5	26
78553	C7988-1420-3B	5340-00-297-0312	4	68
			17	30
0BXW5	ESD5551	6110-01-470-4266	5	60
01276	FA1493FFF3000	4720-00-021-3320	1	22
28818	FF40JM02	5640-01-386-9618	BULK	8
18265	FHG08-0511		7	19
81349	FHN26G1	5920-00-892-9311	5	30
98410	G-775-14	5940-01-369-2271	16	2
			16	6
			16	10
			16	14
			16	18
			16	22
78225	GAR-0661-1	5330-01-465-9273	11	13
98410	H-780-38	5940-01-369-2872	3	2
			3	7
			3	12
			3	17
			3	21
91637	HLM-10-10Z1301J	5905-01-368-2539	5	35
19617	HW310	5310-00-498-7234	2	16
			4	20
			5	5
			5	79
			10	11
			14	6
			14	12
			17	11
			19	7
			25	4
13555	H400	6645-00-089-8842	5	18
81349	JANTX1N1190	5961-00-154-7046	22	11
81349	JANTX1N1190R	5961-01-067-9493	22	1
22175	JM75LC6SS14R	5340-01-470-6235	14	11
50999	JNZ20S0S	2910-01-474-8437	11	67
77342	KUP14D15-24VDC	5945-00-458-3351	5	76
46717	LA-519-9	4730-00-415-3172	19	29
0BXW5	LSS100	5895-01-470-6709	5	64
0BXW5	LSS400	5895-01-470-6702	5	64
31827	MF6T-TPR87-B	5340-01-384-6650	4	124

MARINE CORPS TM 09249A/09246A-24P/3
PART NUMBER INDEX

CAGEC	PART NUMBER	STOCK NUMBER	FIG.	ITEM
81349	MIL-H-13444,TYPE I,1/4 IN. ID.	4720-00-913-5910	BULK	7
06YD3	MSP300-100-P-3-N	6620-01-470-3854	19	6
0BXW5	MSP675C	6695-01-470-3873	19	2
96906	MS16624-3275	5325-01-469-9637	23	7
96906	MS20066-356	5315-00-847-3531	23	6
96906	MS20066-358	5315-01-222-9228	23	5
96906	MS20659-129	5940-00-114-1314	21	31
96906	MS20659-165	5940-00-113-0954	21	46
96906	MS21266-2N	5325-00-074-3301	4	38
96906	MS21306-1G	5310-01-396-5840	2	11
			24	19
			25	7
80205	MS21318-21		21	26
96906	MS24524-22	5930-00-655-4247	5	14
96906	MS25036-110	5940-00-143-4793	19	34
96906	MS25036-128	5940-00-113-9831	14	53
			16	16
			16	20
			16	24
96906	MS25036-154	5940-00-230-0515	22	12
96906	MS25036-157	5940-00-143-4777	23	13
77820	MS25043-18DA	5935-01-470-6221	14	48
96906	MS25224-1	5930-00-615-6731	5	17
96906	MS25281-4	5340-00-988-3210	22	16
96906	MS27183-10	5310-00-809-4058	1	17
			4	19
			7	3
			10	57
			11	63
			14	39
			17	16
			19	20
96906	MS27183-13	5310-00-087-7493	21	17
96906	MS27183-49	5310-01-333-1883	21	44
96906	MS3102R18-11P	5935-00-852-9611	15	19
96906	MS3367-5-9	5975-00-111-3208	5	80
			6	6
			15	4
96906	MS3368-1-9-A	5975-00-944-1499	6	5
96906	MS35206-203	5305-00-889-3118	22	10
96906	MS35206-248	5305-00-984-6196	22	15
96906	MS35206-287	5305-00-988-1728	21	45
96906	MS35333-38	5310-00-559-0070	22	2
96906	MS35333-40	5310-00-550-1130	22	4
			23	11
96906	MS35338-44	5310-00-582-5965	21	29
96906	MS35338-46	5310-00-637-9541	23	2
96906	MS35489-27		14	21
96906	MS35489-88	5325-00-907-1183	25	5
96906	MS35645-1	2590-00-141-9758	11	12

PART NUMBER INDEX

CAGEC	PART NUMBER	STOCK NUMBER	FIG.	ITEM
96906	MS35649-282	5310-00-934-9757	22	3
96906	MS39347-5	5940-00-237-2703	17	17
96906	MS51412-11	5310-01-242-2679	21	5
96906	MS51412-2	5310-01-234-9416	4	3
			5	3
			5	51
			10	36
			11	29
			14	32
96906	MS51412-25	5310-00-044-6477	10	6
96906	MS51412-5	5310-01-234-9415	4	67
96906	MS51412-8	5310-01-267-1685	11	16
			17	2
96906	MS51412-9	5310-01-266-4641	24	1
96906	MS51521B5	4730-00-821-7917	11	4
96906	MS51846-1	4730-00-230-1996	19	24
96906	MS51846-64	4730-00-196-1991	7	28
96906	MS51937-7	5306-00-017-6143	21	23
96906	MS51967-20	5310-00-763-8920	21	24
96906	MS90725-2		23	12
96906	MS90725-3	5305-00-068-0500	21	37
			22	14
96906	MS90725-5	5305-00-068-0501	23	12
96906	MS90725-6	5305-00-068-0502	21	30
96906	MS90726-60	5305-00-269-2803	21	19
81349	M24243/6-A402H		1	2
			19	13
81349	M5086/2-16-9	6145-00-578-6605	BULK	4
93061	NV108P-4	4820-00-785-8153	10	24
			11	33
45722	P-15121-17	5305-00-211-9344	4	95
45722	P-15121-20	5305-00-036-6972	4	125
45722	P-15121-38	5305-01-247-6829	5	41
45722	P-15121-64	5305-00-191-6226	7	35
			14	20
0BXW5	PCI102	6130-01-470-2451	5	101
06383	PLM2S		15	2
06383	PLT2S	5975-01-128-0390	6	4
			12	2
			15	3
			19	27
06383	PN18-6LF-C	5940-01-425-2020	14	36
18265	PPP20-6605	5340-01-368-5596	7	6
18265	P10-1870		8	4
18265	P10-6593	4820-01-192-7676	8	5
18265	P10-9331-016-700	4720-01-375-1930	7	15
18265	P11-9711		8	3
18265	P18-2059	2940-01-378-1130	8	2
61112	QS-1189		11	64
78280	QS-4-2TC	4810-01-067-8276	13	1
56501	RA25177	5940-01-126-3973	7	33

PART NUMBER INDEX

CAGEC	PART NUMBER	STOCK NUMBER	FIG.	ITEM
56501	RA25177	5940-01-126-3973	7	39
			13	3
56501	RA2573	5940-01-112-9746	7	32
			7	38
			13	2
			15	21
81860	RB-X86	5342-01-366-3361	21	11
18265	RBX00-6379	6620-01-369-0432	7	21
56501	RB25177	5940-01-139-0853	5	62
			6	15
15912	RB2573	5940-01-082-3321	5	63
			6	14
			15	20
			18	3
			18	7
			18	11
			18	15
44655	REL5R0	5905-01-336-7533	5	32
56501	RG9731	5940-00-113-8191	14	51
			16	4
			16	8
			16	12
78276	RNS832-120KT2	5310-01-470-3877	5	42
64533	RP-5	5975-00-879-7234	5	46
81640	SM50D8	5945-01-365-9953	14	19
70411	SP2529VT	4820-01-367-1836	19	26
28105	ST-301-1RED	5970-01-470-1631	3	3
			3	8
			3	15
			3	23
28105	ST-301-1WHITE	5970-01-470-1637	3	5
			3	10
			3	14
			3	19
			3	24
50019	S1990	4730-00-595-3108	7	26
0BXW5	TCM100-50/60HZ	5998-01-466-4724	5	58
0BXW5	TCM102	5998-01-466-4726	5	54
0BXW5	TCM400-400HZ	5998-01-466-6097	5	58
83616	TD2-03420	4720-00-804-9249	11	68
1E045	TL62B7X1/16	9390-01-470-1205	10	61
30554	TYPE II GRADE A		4	21
66836	T04045TF151	2815-01-462-2289	24	12
93061	V500P-12	4820-01-008-2922	20	6
76385	Z-3170		14	9
1X968	ZX-4134	5330-01-369-8947	BULK	6
4T483	00.625TF12-50	4720-01-470-4779	BULK	11
79470	00904F-504-J04-0 0963	4720-01-386-1856	11	7
1FP59	012316-7	5310-00-761-6882	21	43
			22	13

PART NUMBER INDEX

CAGEC	PART NUMBER	STOCK NUMBER	FIG.	ITEM
27264	02-09-1104	5999-01-039-8438	15	10
1DG36	021034	2990-01-470-7119	7	9
27264	03-09-1042	5935-00-315-9563	15	9
27264	03-09-2022	5935-00-482-7721	12	4
71600	1-3-0		4	64
5G996	1/0SGT	6145-01-470-6195	BULK	2
25795	1R424A	2940-01-470-6444	9	1
77820	10-40450-18	5330-00-290-9891	5	47
77820	10-40450-22	5330-00-641-4336	5	50
30780	10M18C80MX	4730-01-470-3971	20	1
36378	1001547-01	4730-01-134-9827	11	18
79470	1069X6	4730-00-041-2526	11	54
30327	112B1-4IN	4730-00-765-9103	11	39
18310	1127-38-0516	5940-00-954-3558	5	29
19207	11608950-18	4730-01-296-9318	11	19
19207	11674728	5935-01-097-9974	2	17
04826	12-130109G		20	8
81495	1223-299	5920-01-459-3847	5	31
19207	12325869	5306-01-156-7663	2	18
			4	25
			5	4
			5	81
			10	14
			14	3
			14	10
			17	8
19207	12412364	5930-01-372-5387	5	12
15526	125AZY-10M	5310-01-470-1373	10	44
			19	16
			21	1
15526	125AZY-16M	5310-01-470-1381	24	6
			24	8
15526	125AZY-5M	5310-01-470-1370	19	31
93061	125HBL-10-12	4730-01-340-7594	20	5
93061	125HBL-4-4	4730-00-200-0531	11	56
15526	127BZY-10M	5310-01-470-2355	21	2
15526	127BZY-12M	5310-01-470-0450	24	9
15526	127BZY-16M	5310-01-470-0455	24	5
15526	127BZY-5M	5310-01-470-0446	19	32
15526	127BZY-8M	5310-01-470-0448	10	53
06555	127678	2910-01-366-8983	11	30
97403	13230E6744-62	5310-01-469-9836	4	2
			5	2
			5	52
			7	34
			11	28
97403	13230E6744-63	5310-01-469-9864	4	18
			5	95
			7	4
			10	58
			11	22

PART NUMBER INDEX

CAGEC	PART NUMBER	STOCK NUMBER	FIG.	ITEM
97403	13230E6744-63	5310-01-469-9864	14	42
			17	15
			19	19
97403	13230E6744-64	5310-01-469-9865	10	7
			11	59
97403	13230E6744-65	5310-01-469-9866	2	15
			7	17
			10	3
			17	29
			25	8
52152	1430		BULK	12
1JB11	15300027	5935-01-470-7643	15	8
98441	160-12	4720-01-470-3917	BULK	10
5Y407	1792346		6	1
5Y407	1792391		6	10
79470	1908	4730-01-387-4000	10	27
60177	19390	5950-01-368-2915	14	34
79470	1945	4730-01-414-4281	10	28
81343	2-4-070202	4730-01-470-6338	19	28
91929	2TL1-3	5930-01-453-2655	5	16
16327	2Z329B		7	25
77342	20C318	5340-01-370-8349	5	69
30327	204SAE	4820-00-289-3378	19	3
5U990	205-605	4730-01-407-0649	11	3
98441	208-4	4720-01-470-3929	BULK	3
98441	208-5	4720-01-470-6230	BULK	5
08928	21NTE102	5310-01-406-1672	4	106
			5	55
08928	21NTE616	5310-01-366-4412	17	34
08928	21NTE813	5310-01-365-5788	17	1
019L2	21NTM82	5310-01-396-5836	10	50
			19	35
28520	2314	4730-01-378-5224	10	26
28520	2326	4730-01-370-5426	10	34
			11	35
02768	236-170406-04	5320-01-366-4394	4	8
77342	24A071	5340-01-078-9038	5	99
77342	27E893	5935-01-367-4422	5	68
24617	274825	5305-01-056-1501	4	56
			7	5
70485	2774	5325-00-319-0547	25	1
60177	29400-2	5925-01-466-8153	5	23
41197	3A43788B		10	12
98441	30682-10-10B	4730-00-234-7637	20	2
45454	320777	2930-01-366-6107	10	54
9R803	330JS	5999-01-366-2621	5	71
9R803	3300-10-XP-74	5940-01-470-3031	14	14
9R803	3300-2	5940-01-470-2470	5	73
9R803	3300-2-XP-74		5	72
9R803	3300-4	5940-01-470-2875	5	66
9R803	3300-4-XP-74		5	65

PART NUMBER INDEX

CAGEC	PART NUMBER	STOCK NUMBER	FIG.	ITEM
63123	35-000252-001	5999-01-092-2655	12	3
79260	35794	5340-01-223-4373	7	8
98441	38300-1-56-48RE	4720-01-369-9366	7	11
81343	4-2-140137		11	17
81343	4-4-070102	4730-01-470-6039	11	47
81343	4-4-4-140424	4730-01-470-6199	11	53
81343	4-5-070102		11	34
81483	40-1389	5905-01-063-9644	17	26
77342	40G432	5970-01-280-0362	5	98
1DG36	400112	1560-01-470-2391	7	1
72850	40194	2910-01-366-7293	12	1
21450	444069	4730-00-613-6468	11	40
83330	47-0901-2900-201	6210-00-935-6919	5	9
72850	479735	2940-01-365-6535	11	48
81343	5-5-070118	5310-01-475-2155	11	24
81343	5-5-070601	4730-01-470-6041	11	5
7T184	5W133	5905-01-470-3203	5	33
78189	501-040800-00	5310-00-836-3520	5	34
			14	46
78189	501P06080000AMB	5310-00-063-7360	4	96
			5	21
78189	511-081800-00	5310-00-052-3632	5	37
			11	25
			14	16
1EG71	521-9181	5980-01-074-1633	5	10
32529	53	4030-00-272-9002	10	13
99739	55-1986-19	4720-01-470-5994	BULK	1
22175	55MC1-11-SS-R	5340-01-470-6499	25	3
5Y407	5600014		6	2
5Y407	5600030		6	3
77824	5798-190Z-3-75.0	5330-01-406-5707	14	4
71744	6S6/30V-801	6240-01-470-4272	5	8
3Z031	602350050	4730-01-470-5158	7	27
78189	61-101041-90-014 2B-0542B	5305-01-187-5878	14	33
06324	660-005C18S4.5-0 1	5935-01-470-6221	5	44
93742	69-539-2	4730-00-812-1333	11	23
30554	69-583		4	62
30554	69-662-36	5305-00-036-6902	5	70
			14	13
30554	69-662-5	5305-00-224-1092	5	36
			14	47
36156	702807-01		21	20
36156	703512-01		21	40
36156	707509-01		23	4
36156	707807-02		23	9
36156	716511-02		23	3
36156	718514-01		21	38
36156	718515-01		21	39
36156	718516-01		21	42

PART NUMBER INDEX

CAGEC	PART NUMBER	STOCK NUMBER	FIG.	ITEM
36156	718517-01		21	41
30554	72-2098-2		14	43
30554	72-2236		17	21
30554	72-5304	4730-01-470-1982	20	7
36156	720510-0B		21	32
36156	720982-0A		21	32
36156	725819-0A		21	28
36156	777056-0A	2920-01-298-6321	23	10
36156	777066-0A	2920-01-467-2624	21	21
36156	777067-0A	2920-01-467-2626	21	21
36156	778663-0B		21	15
31874	7812-01		10	55
36156	789293-0A	6115-01-385-3104	21	33
36156	791150-0A		23	8
83330	800-1030-0337-50 4	6210-00-583-9349	5	7
0UJ55	800S	6140-01-378-8232	2	7
36156	801048-04		21	36
70485	804	5325-00-174-9038	10	19
02768	8070-25-00	5310-01-407-7191	4	76
36156	832805-01		23	7
36156	834822-01		21	27
94222	85-12-460-16	5325-01-161-2654	5	90
94222	85-16-400-16	5307-01-374-4451	4	58
94222	85-34-101-20	5310-00-822-8525	4	61
			5	89
94222	85-35-309-56	5325-01-301-7903	4	60
			5	84
94222	85-46-103-39	5310-01-365-4381	5	88
81640	8530K9	5930-01-398-6692	5	27
36156	865873-01	5330-01-369-7318	21	35
8W764	8719	6145-01-470-6341	15	7
30554	88-20013		4	51
30554	88-20014		4	48
30554	88-20049	4930-01-475-0388	11	52
30554	88-20064-7		21	25
30554	88-20073		1	9
30554	88-20074		1	10
30554	88-20075		1	19
30554	88-20102	9905-01-377-5094	1	6
30554	88-20110		1	12
30554	88-20123		4	100
30554	88-20126		1	13
30554	88-20188	5999-01-382-8223	2	3
30554	88-20218	2920-01-388-2776	4	65
30554	88-20247-2	5310-01-365-3190	24	16
30554	88-20247-3	5310-01-365-8139	21	14
30554	88-20258	5340-01-470-2973	5	97
30554	88-20260-11	5305-01-470-1425	5	39
30554	88-20260-2		4	1
			19	10

PART NUMBER INDEX

CAGEC	PART NUMBER	STOCK NUMBER	FIG.	ITEM
30554	88-20260-20	5306-01-368-8041	4	50
30554	88-20260-23	5305-01-365-6313	4	4
			5	1
			11	15
			25	2
30554	88-20260-25	5305-01-365-6314	17	13
30554	88-20260-30	5305-01-369-2166	4	17
			19	22
30554	88-20260-31	5306-00-484-5730	4	40
			5	94
			11	21
30554	88-20260-33	5306-01-366-7075	4	66
30554	88-20263	5905-01-470-1376	5	74
30554	88-20286	5330-01-366-2836	11	31
30554	88-20305-1		17	24
			18	1
30554	88-20305-2		17	25
			18	5
30554	88-20305-3		17	23
			18	9
30554	88-20305-5		17	3
			18	13
30554	88-20450-1		14	52
30554	88-20450-4		6	16
			15	22
			18	2
			18	6
			18	10
			18	14
30554	88-20461		4	104
30554	88-20468	2930-01-370-2868	10	64
30554	88-20470		15	5
30554	88-20540-11		16	3
			16	7
			16	11
			16	15
			16	19
			16	23
30554	88-20544-4	5340-01-470-6228	19	8
30554	88-20546-4	5340-01-470-6324	5	78
			19	17
30554	88-20555-2	5305-01-470-1421	5	91
30554	88-20561-1	4730-01-470-1626	10	17
			11	2
30554	88-20561-2	4730-01-470-2409	7	14
			20	3
30554	88-20561-3	4730-01-470-1567	7	22
30554	88-20561-4	4730-01-470-1595	10	9
30554	88-20561-6	4730-01-470-1578	7	2
			11	51
30554	88-20561-7	4730-01-470-1701	11	26

PART NUMBER INDEX

CAGEC	PART NUMBER	STOCK NUMBER	FIG.	ITEM
30554	88-20564-12	5310-00-044-6477	11	58
30554	88-20564-14	5310-01-466-4926	17	19
30554	88-20565-1	5305-01-470-4533	10	46
30554	88-20565-2	5305-01-470-5134	5	93
30554	88-20568-1	5310-01-470-1981	1	18
			14	45
30554	88-20568-2	5310-01-470-1998	2	9
30554	88-20568-4	5310-01-470-6529	24	18
30554	88-20568-5	5310-01-470-2776	24	17
30554	88-20568-6	5310-01-470-3024	21	4
38151	88-21007	6115-01-395-6536	21	10
30554	88-21007-36		21	47
38151	88-21008	6115-01-387-4676	21	10
30554	88-21015	5950-01-382-3371	14	37
30554	88-21044	2910-01-377-3984	11	57
30554	88-21046		8	1
30554	88-21098		4	39
30554	88-21099	5340-01-467-0760	4	43
30554	88-21147	5120-01-375-4373	17	9
30554	88-21182-2	5940-01-470-2768	14	15
30554	88-21585		4	118
30554	88-21588		4	88
30554	88-21589		4	92
30554	88-21594		4	9
30554	88-21595		4	37
30554	88-21603		1	23
30554	88-21634		1	5
30554	88-21649-19		22	6
30554	88-21649-87		22	9
30554	88-21667		17	33
30554	88-21669		14	40
30554	88-21680		4	87
30554	88-21681		4	79
30554	88-21683		2	10
30554	88-21684		2	12
30554	88-21685		2	8
30554	88-21711		10	2
30554	88-21731		25	10
30554	88-21732		4	59
30544	88-21733		4	57
30554	88-21743		7	18
30554	88-21755-2	4730-01-301-4257	7	30
30554	88-21767		17	32
30554	88-21770		4	71
30554	88-21771		4	117
30554	88-21775		17	27
30554	88-21776		19	12
30554	88-21813		4	101
30554	88-21814		4	98
30554	88-21815		4	97
30554	88-21820		4	116

PART NUMBER INDEX

CAGEC	PART NUMBER	STOCK NUMBER	FIG.	ITEM
30554	88-21851	4710-01-470-2119	11	41
30554	88-21854		11	61
30554	88-21870		4	123
30554	88-21874		5	104
30554	88-21875		4	119
30554	88-21877		4	120
30554	88-21879	4140-01-467-1386	10	51
30554	88-21882		24	15
30554	88-21883		11	44
30554	88-21889		4	90
30554	88-21892		11	11
30554	88-21893	5340-01-465-7765	11	14
30554	88-21908	4720-01-373-0526	BULK	9
30554	88-21914		4	74
30554	88-21922		21	15
30554	88-21926		4	44
30554	88-21927		4	115
30554	88-21932		14	8
30554	88-21933		17	12
30554	88-21951		4	85
30554	88-21952		2	19
30554	88-21953		4	80
30554	88-21963		4	73
30554	88-21964		4	63
30554	88-21968		7	12
30554	88-21973		4	83
30554	88-21977		4	55
30554	88-21987		4	27
30554	88-21988		4	16
30554	88-21998		4	49
30554	88-22008		1	24
30554	88-22020		24	7
30554	88-22031		4	89
30554	88-22037		4	72
30554	88-22039		14	22
30554	88-22040		4	70
30554	88-22049		10	37
30554	88-22050		4	54
30554	88-22058		4	69
30554	88-22063		4	103
30554	88-22068	4720-01-392-0319	11	50
30554	88-22073		4	52
30554	88-22084		4	82
30554	88-22092		4	24
30554	88-22111		11	6
30554	88-22117		24	13
30554	88-22120		5	82
30554	88-22126-1		14	29
			16	1
30554	88-22126-2		14	30

PART NUMBER INDEX

CAGEC	PART NUMBER	STOCK NUMBER	FIG.	ITEM
30554	88-22126-3		14	31
			16	9
30554	88-22126-4		14	50
30554	88-22126-5		14	27
			16	13
30554	88-22126-6		14	26
			16	17
30554	88-22126-7		14	25
			16	21
30554	88-22130		24	14
30554	88-22136		17	10
30554	88-22145		10	1
30554	88-22146		17	20
30554	88-22161-2		19	21
30554	88-22162		19	23
30554	88-22163-1		17	5
30554	88-22205		21	18
30554	88-22209	6150-01-406-9533	1	21
30554	88-22318		17	28
30554	88-22418-1	5961-01-431-4558	5	28
30554	88-22418-2	5961-01-470-4673	14	24
30554	88-22429	2990-01-389-3003	4	6
30554	88-22430	5340-01-367-8956	4	7
30554	88-22437	5940-01-384-0384	14	41
30554	88-22469	4020-01-470-6597	17	7
30554	88-22472		4	46
30554	88-22473		4	86
30554	88-22474		4	114
30554	88-22475		4	111
30554	88-22476		4	113
30554	88-22478		4	109
30554	88-22479		4	108
30554	88-22481		19	25
30554	88-22482	5365-01-431-4603	4	110
30554	88-22483	5365-01-431-4540	4	105
30554	88-22500		4	45
30554	88-22501		4	47
30554	88-22509	6110-01-372-2597	1	16
			14	44
30554	88-22526		21	9
30554	88-22527		21	8
30554	88-22530	5305-01-366-3501	21	6
30554	88-22546	2910-01-470-7423	11	8
30554	88-22553	5945-01-470-0491	11	65
30554	88-22582		4	26
30554	88-22583		4	14
30554	88-22584		4	10
30554	88-22585		4	11
30554	88-22586		4	15
30554	88-22587		4	13
30554	88-22588		4	12

PART NUMBER INDEX

CAGEC	PART NUMBER	STOCK NUMBER	FIG.	ITEM
30554	88-22591		4	77
30554	88-22592		4	53
30554	88-22593		4	84
30554	88-22599		4	35
30554	88-22600		4	31
30554	88-22666		4	102
30554	88-22701		14	1
30554	88-22702		4	34
30554	88-22703		10	25
30554	88-22704		14	2
30554	88-22705		4	28
			14	5
30554	88-22712	5330-01-470-4206	10	48
30554	88-22717		10	47
30554	88-22718		4	75
30554	88-22719		10	62
30554	88-22720		10	49
30554	88-22723		4	23
30554	88-22724		4	78
30554	88-22730		4	81
30554	88-22737		1	9
30554	88-22738		1	14
30554	88-22758	5910-01-384-1745	14	49
30554	88-22761	5915-01-396-9253	17	22
30554	88-22762	5915-01-394-0942	17	6
30554	88-22774		5	83
30554	88-22786	5310-01-466-7247	17	18
30554	88-22790-1	5310-01-466-6312	4	41
			7	7
			10	66
			11	49
			19	18
			21	7
30554	88-22790-2	5310-01-470-1286	10	8
			11	60
30554	88-22790-3	5310-01-466-6687	7	16
			10	35
			25	9
30554	88-22791-1	5305-01-466-9756	4	36
30554	88-22791-2		4	99
		5305-01-467-1561	5	22
30554	88-22793-4	5305-01-464-6667	4	42
			5	24
30554	88-22793-6	5305-01-470-2911	5	77
30554	88-22793-7	5305-01-470-2429	4	112
30554	88-22793-8	5305-01-470-2440	4	107
30554	88-22793-9	5305-01-470-2448	11	27
30554	88-28815		4	118
81640	8906K4522	5930-01-365-9614	19	11
81640	8906K4523	5930-01-376-2001	5	11
			5	13

PART NUMBER INDEX

CAGEC	PART NUMBER	STOCK NUMBER	FIG.	ITEM
81640	8906K4750	5930-01-368-2892	5	19
48370	933-M10	5306-01-470-2097	10	42
			19	14
			21	3
48370	933-M12	5306-01-470-3069	24	10
15526	933-ZY-M16-50	5306-01-470-1976	24	4
15526	933-ZY-M8-65	5306-01-470-3057	10	52
81640	9565H150	6110-01-367-8921	14	28
30554	96-23500		1	1
30554	96-23501		1	1
30554	96-23504	6110-01-466-4725	5	6
30554	96-23505	6110-01-466-4723	5	6
30554	96-23506-01		1	4
30554	96-23506-02		1	4
30554	96-23508		5	87
30554	96-23509		1	20
30554	96-23510		1	7
30554	96-23512		1	8
30554	96-23513		1	8
30554	96-23516		5	85
30554	96-23517		5	25
30554	96-23518		5	96
30554	96-23519		5	92
30554	96-23521		4	32
30554	96-23522		24	11
30554	96-23524		4	30
30554	96-23525		4	33
30554	96-23527		4	22
30554	96-23528		24	3
30554	96-23529		10	67
30554	96-23534	5930-01-470-8799	19	1
30554	96-23542	6110-01-470-4263	5	61
0FNW8	96-23545	6110-01-470-4245	5	38
30554	96-23546	6150-01-470-6001	15	11
30554	96-23547	2930-01-471-5127	10	38
30554	96-23550	4720-01-470-3016	10	21
30554	96-23551	4720-01-470-2086	10	10
30554	96-23555		19	9
30554	96-23556		5	107
30554	96-23558	4140-01-470-6558	10	40
30554	96-23559		10	56
30554	96-23561		2	5
			3	20
30554	96-23562		2	6
			3	11
30554	96-23563		2	1
			3	1
30554	96-23564		2	13
			3	16
30554	96-23565	6150-01-470-6017	5	48
30554	96-23566		2	20

PART NUMBER INDEX

CAGEC	PART NUMBER	STOCK NUMBER	FIG.	ITEM
30554	96-23566		3	6
0FNW8	96-23569	7025-01-467-4709	5	40
30554	96-23571-01		1	3
30554	96-23571-02		1	3
0FNW8	96-23574	6150-01-472-8402	5	75
30554	96-23575		14	23
30554	96-23577		11	9
30554	96-23578		11	55
30554	96-23581-2	4710-01-470-0876	11	32
30554	96-23582-1	4140-01-470-6555	10	60
30554	96-23588	5340-01-470-6357	7	10
30554	96-23593-2		11	42
30554	96-23595	4720-01-470-3561	10	23
30554	96-23596		1	15
30554	96-23599		10	20
30554	96-23600		14	7
30554	96-23602		10	63
30554	96-23604		1	11
30554	96-23605		10	41
30554	96-23607	5342-01-367-1516	24	20
30554	96-23609-2		10	31
30554	96-23609-5		10	30
			11	36
30554	96-23610-1		10	18
30554	96-23610-2		10	29
30554	96-23610-5		10	65
30554	96-23610-7		10	32
30554	96-23613-2	5310-01-470-2355	10	43
			19	15
30554	96-23614-1		20	4
30554	96-23616	5935-01-470-7699	5	105
30554	96-23617		4	29
30554	96-23634		5	106
30554	96-23635		5	103
30554	96-23636		19	5
30554	96-23637		3	4
			3	9
			3	13
			3	18
			3	22
30554	96-23640	4730-01-470-3579	19	4
30554	96-23650		6	11
30554	96-23651		6	12
30554	96-23652	4730-01-470-3566	10	22
30554	96-23654		2	21
0FNW8	96-23664	5995-01-470-2882	5	57
0FNW8	96-23665	5995-01-470-2881	5	56
30554	96-23666	5905-01-470-0564	5	67
30554	96-23671		19	36
30554	96-23672		7	24
30554	96-23673		5	86

PART NUMBER INDEX

CAGEC	PART NUMBER	STOCK NUMBER	FIG.	ITEM
30554	96-23678		14	18
30554	96-23680-1		15	12
30554	96-23681-1		7	29
30554	96-23681-2		7	13
30554	96-23683	5995-01-470-4269	5	49
30554	96-23685		5	100
30554	96-23687		10	45
30554	96-23690-1		11	20
30554	96-23690-10		11	45
30554	96-23690-3		11	43
98441	96-23690-4		11	1
30554	96-23690-5		11	38
30554	96-23690-7		10	39
			11	46
30554	96-23697		7	37
30554	96-23700-2		7	36
30554	96-23702-2		7	40
30554	96-23703		7	31
30554	96-23705-3		7	23
30554	96-23710	5342-01-470-5177	10	16
77820	97-3108B-28-15P		15	1
77820	97-3108B-28-15S		15	13
94222	97-50-170-11		4	94
02660	9760-22	5935-01-005-9478	5	43